Advanced Technologies and Societal Change

More information about this series at http://www.springer.com/series/10038

Reiner Wichert · Beate Mand
Editors

Ambient Assisted Living

9. AAL-Kongress, Frankfurt/M, Germany, April 20–21, 2016

Editors
Reiner Wichert
SageLiving GmbH
Pfungstadt
Germany

Beate Mand
Verband der Elektrotechnik Elektronik
Informationstechnik e.V.
Frankfurt
Germany

ISSN 2191-6853 ISSN 2191-6861 (electronic)
Advanced Technologies and Societal Change
ISBN 978-3-319-84876-1 ISBN 978-3-319-52322-4 (eBook)
DOI 10.1007/978-3-319-52322-4

Softcover reprint of the hardcover 1st edition 2017

Printed on acid-free paper

This Springer imprint is published by Springer Nature
The registered company is Springer International Publishing AG
The registered company address is: Gewerbestrasse 11, 6330 Cham, Switzerland

Preface

The habitats of tomorrow will be connected through networked, autonomous, and assistive systems. Houses, apartments, offices, transport, or public spaces are converting into health places. Assistance technologies enable a smooth transition from comfortable health support to medical or nursing care. To enable this, Active Assisted Living (AAL) combines a whole range of innovative key technologies from these domains. Nowadays, we can recognize the trend that precisely this community is discussing domain spanning system concepts to integrate seamlessly and spontaneously the various components and solutions into an overall system approach.

While this potential has been recognized for some time, breakthroughs in terms of widespread availability and deployment of solutions have yet to be achieved. The EU and the AAL Association have funded activities in this area for some years, and some of these are now at a stage in their development where direct hands-on involvement of development companies is the best way to make sure that this work produces results that are effective and applicable in real industrial settings.

To follow these goals, a conference series has been established as an annual showcase event for the people involved in this community: the AAL-Kongress (Congress for Active Assisted Living) with its purpose is to exhibit and demonstrate ICT solutions, promote networking within the community, provoke debate on various topics and highlight new or emerging developments in the area to inform the AAL community and discuss the problems and challenges we have to face in the common years.

The first AAL Kongress 2008 had the focus on applications of intelligent assistive systems within the areas of "health & homecare", "safety & privacy", "maintenance & housework" and "social environment", At the second AAL-Kongress, more than 520 participants attended. It focused on use cases to support the manufacturing of products adjusted to the needs of the user. In 2010 the third AAL-Kongress had been organized with close to 600 participants also with the focus on use cases. In 2011, it advanced to the leading congress for AAL with 870 participants. In 2012, the focus was laid on technologies in a self-determined life and the number of participants passed over 1000, still addressing economic challenges and trendsetting applications on innovative technology. In 2013, the sixth AAL-Kongress was focussing on "quality of life in times of changing demography and technology". Within the thematic topic "Better Life with Assistive Technologies" the congress addressed in 2014 the basic human needs in the different areas of housing, mobility, work, health and care. In 2015, the format of the conference series has been changed by combining the AAL-Kongress with the fair "Zukunft Lebensräume" to bring AAL closer to the people.

In 2016, the series continued in the combined format and with 750 participants the congress was again a great success with its excellent platform to exchange knowledge between all stakeholders, from developers, manufacturers and users, service providers, end users and representatives from politics, industry and associations. This time the congress focussed on technology for assistance in health, independence and comfort. Close to 500 authors from six different countries had submitted contributions where 108 papers had been accepted. After a solid review process, 14 papers were accepted to be included in these scientific proceedings of the conference. Three independent reviewers were matched by their expertise area to the topic of each paper.

In closing, I would like to thank the 36 reviewers of the Reviewing Committee, all the authors, the organizers of this event and the conference participants who helped to make this congress a success.

Reiner Wichert
Chair of Scientific Program Committee
SageLiving GmbH
Pfungstadt, Germany

Organizing Committee

Chair Scientific Program Committee

Reiner Wichert, SageLiving GmbH

Scientific Program Committee/Review Committee 9. AAL Kongress "Zukunft Lebensräume Kongress 2016"

Jan Alexandersson, DFKI Saarbrücken
Rashid Asarnusch, Zentrum für Telemedizin Bad Kissingen
Serge Autexier, DFKI Bremen
Daniel Bieber, Institut für Sozialforschung und Sozialwirtschaft
Michael Brach, Universität Münster
Martin Braecklein, Linde Healthcare
Andreas Braun, Fraunhofer IGD
Alexandra Brylok, VSWG
Wolfgang Deiters, Fraunhofer ISST
Marco Eichelberg, Offis
Uwe Fachinger, Universität Vechta
Melina Frenken, Jade Hochschule
Petra Friedrich, HS Kempten
Birgit Graf, Fraunhofer IPA
Ingrid Hastedt, Wohlfahrtswerk Baden-Württemberg
Andreas Hein, Universität Oldenburg
Svenja Helten, Universität Vechta
Oliver Koch, Hochschule Ruhr West
Benno Kotterba, iAQ Institut für Assistenzsysteme und Qualifizierung e.V.
Petra Knaup-Gregori, Universität Heidelberg
Harald Künemund, Universität Vechta

Contents

Part I
Technical Assistance for Urban Areas

Integration of Stationary and Wearable Support Services for an Actively Assisted Living of Elderly People: Capabilities, Achievements, Limitations, Prospects—A Case Study

Rainer Lutze and Klemens Waldhör

Abstract Within the recent three years, a stationary home assistance system has been developed, continuously optimized and operated for supporting seniors of very high age. In the last year, the scope of the system has been extended by function and beyond the spatial borders of the familiar home by a smartwatch with integrated cellular radio (Samsung Gear™ S) as a wearable device. All condensed data from the different stationary and mobile sensors are transferred to and collected by a central server for long-term analysis. The technical structure of the system is presented and its capabilities will be described, especially with respect to the variation of collected data over time in the course of a progressing dementia of one of the inhabitants. The different achievements and perceived value, which the system delivers to its users and their relatives over the course of the years will be presented. But also the limitations of the currently available technology in comparison the actual demand of the inhabitants and their relatives will be characterized which defines the boundary conditions and guidelines for further research.

Keywords Wearables · Smartwatches · Stationary home assistance systems · Actively assisted living (AAL) · Sensor fusion · Long-term analysis · Elderly people · MCI · Dementia

1 Problem Description

For most elderly people staying in their familiar own home as long as possible in a self-determined and safe manner way is a very high-ranking target. But on the other hand technical support systems for an assisted active living (AAL) have not gained

R. Lutze (✉)
Dr.-Ing. Rainer Lutze Consulting, D90579 Langenzenn, Germany
e-mail: rainer.lutze@lustcon.eu

K. Waldhör
FOM University of Applied Sciences, D45127 Essen/Nuremberg, Germany
e-mail: klemens.waldhoer@fom.de

R. Wichert and B. Mand (eds.), *Ambient Assisted Living*,
Advanced Technologies and Societal Change,
DOI 10.1007/978-3-319-52322-4_1

a broad acceptance among the elderly. In our case study work we describe (a) the technical assistance solution implemented and its delivered benefits against several concerns of the elderly and (b) the actual usage and perceived benefits for a senior couple of high age in a typical care-giving situation where one spouse gives care to the other in the presence of beginning and progressing (vascular) dementia. The duration of our study covers a service period of nearly three years, for which the assistance system with its increasing functional coverage has been in uninterrupted 24/7 operations.

We have chosen the case study method because a key goal of our research was to perform an in-depth analysis of the usage of such a system in a real live scenario for several years thus combining both a qualitative and quantitative approach. A main concern about applying AAL technologies (cf. [1]) deals with not really involving the elderly, but just focusing on the technical implications rather than including also the psychological and social aspects of such AAL systems. The relatively long study period of nearly three years—including the chance to investigate the upcoming and integration of a new technology like smartwatches—gave us the unique opportunity to get new insights into this difficult area of applying new technological innovations. Thus the study is based both on qualitative data derived from individual experiences and talks with the involved parties and quantitative data derived from the technological systems used, also combining both data when necessary.

In order to monitor the wellbeing of persons in need of support, care, for their daily living and health we use an established technical approach, the detection of the activities and events of daily living (ADLs, EDLs) (see Sect. 3.1 for details).[1] EDLs may be »*falls*« or events like »*retiring to bed at nighttime*« or »*getting up in the morning*«, which determine the beginning, ending and duration of the ADL »*bedtime*« rsp. »*nightly sleep*«. The ADL and EDL detection is performed via stationary sensors deployed within the home or sensors, smartwatches, worn by the residents on their wrist. ADL/EDL detection typically includes sensor fusion from multiple sensor devices. After possibly combining matching EDLs into single ADLs, the duration, presence and intervals between ADLs over time is analyzed. These parameters are combined in a wellbeing function w(t) (see Sect. 3.2 below for details), which characterizes the assumed wellbeing resp. health state of a monitored person at a specific time t. Whenever w(t) falls below a predefined threshold, a *health hazard alert* will be issued. Before external help will be called, the person in need of support resp. their present relatives or caregivers always receive a local pre-alert giving them the opportunity to cancel the upcoming external alert and to filter out occasional false alerts. External help can be provided or organized by distant family members or home emergency call centers (HECCs). In order to clarify the specific hazardous situation, external alert handling always

[1]ADLs have been a central issue in organizing professional nursing practice and for determining the independency status of elderly people, they have been introduced by Sidney Katz more than 60 years ago. In Germany, Liliane Juchli has elaborated these ADLs for a systematic professional care management [2]. In our work, we focus on a small subset of computationally tractable ADLs.

includes establishing a direct speech connection to the home resp. the person in need of support.

The acceptance barriers of elderly people against the utilization of such assistance systems have multiple aspects:

- **Fear of stigmatization**. Visible AAL systems clearly demonstrate to outsiders, but also for the users themselves, that the user indeed needs support for organizing the daily living, a situation that typically everyone want to camouflage as long as possible. A potential solution is the use of "dual use" devices like smartwatches, which cannot be easily identified in their additional assistive usage.
- **Resistance to change**, especially if the installation of the assistance system at home is combined with demolition work. This general human attitude grows with increasing age and can only be compensated by minimizing the necessary construction work for the assistance system.
- **Privacy concerns**. Of course, the continuous monitoring of one's daily living by an assistance system produces a constant feeling of discomfort and raises concerns about the potential misusage of the accumulated data. Especially imaging technologies—despite their technological achievements—are rejected.
- **Costs**. High installation costs, esp. for stationary systems, are another important key point for not using assistance systems. For a broad acceptance and mass distribution of such assistance systems, an acceptable level of assistance services has to be provided at reasonable costs. [4] reports that the average costs Germans would accept are around 20 € service costs per month. In [5] this cost estimation is also confirmed for the US with monthly costs of US $25 reported for the California based *Lively* service (with initial setup costs of US $40).

These aspects all condense in the self-insight that with the usage of the assistance system the *final phase* in one's life has begun, foreseeable followed by the *death*. People tend to delay this type of introspection as long as possible. Thus these technological solutions are mainly perceived through these negative connotations, not by the positive aspects associated. It has been argued [6] that the user concerns can be alleviated and a buying decision for such a technology can be boosted by stressing the **multivalent utility** of the assistance system not only for *support services*, but also for improving *comfort* and *safety* at home as well as for improving the *energy efficiency* of the home (by reduced heating, cooling costs).

2 System Structure

The system developed over the course of the years consists of the following three components (see Fig. 1):

1. A **stationary assistance subsystem** in the home based on high quality presence sensors—multisector PIR sensors—in each room (cf. [7] for a detailed

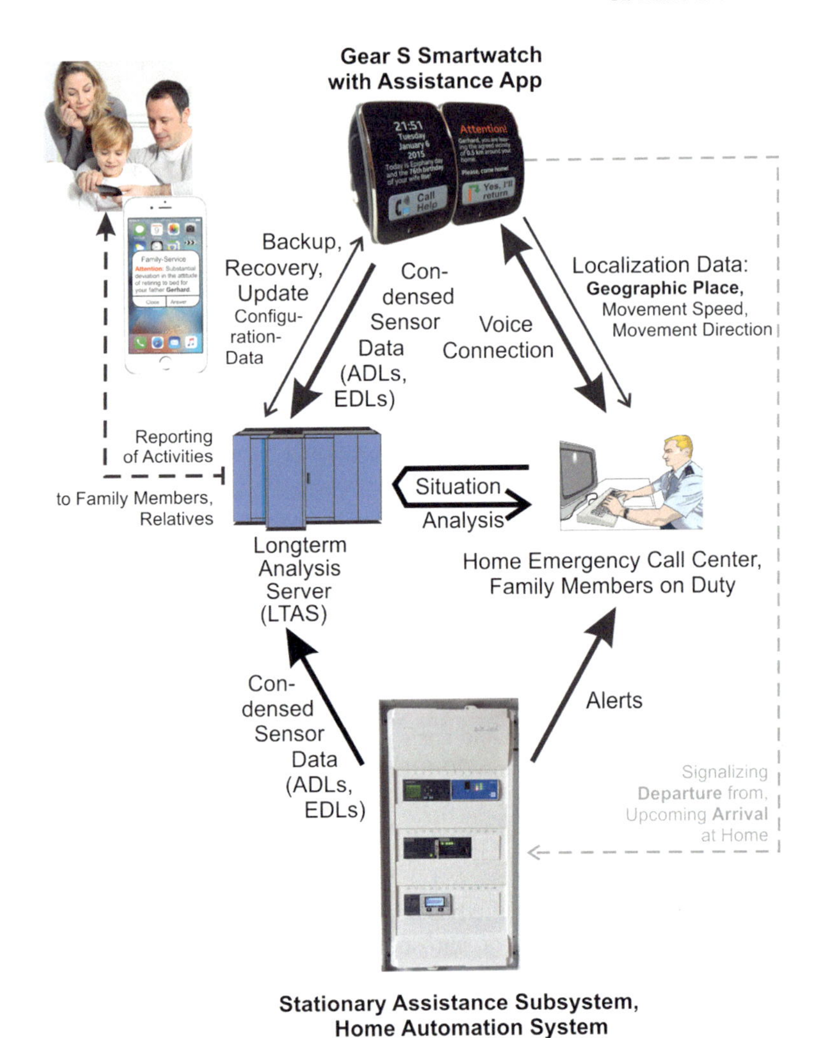

Fig. 1 System structure with components: 1. Stationary assistance subsystem (*bottom*), 2. Smartwatches (*top*) and 3. LTAS (*middle left*). Involved in the communication processes, but not part of the assistance system, is the home emergency call center or family members on duty (*middle right*), which react to alerts and smartphones (of family members, relatives) as endpoints of LTAS services (*upper left corner*)

description, Figs. 2, 3). Sensor fusion and the local monitoring is performed by a Siemens LOGO™ SPS/PLC. Reporting of detected EDLs, ADLs to the LTAS server (via http) and alerting (via E-Mail, SMS) is done in a highly availably

Fig. 2 Two channel presence sensors used for the stationary assistance subsystem (Theben Office™)—1st channel for (local) automatic room lighting control, 2nd channel for presence signaling to assistance system

way via WAN and also cellular network by an INSYS IMO-1™ GRPS router/rule-based fault transmitter. The assistance subsystem simultaneously also acts as the center of a local home automation system (Fig. 4).

2. Smartwatches as **wearables devices** together with our developed assistance app on the wrist of the person in need of support for their daily living/health. Here we use the Samsung Gear™ S smartwatch with its large 2'' AMOLED display, GPS and integrated cellular radio, which can operate independently from a coupled smartphone and is equipped with our Tizen™ assistance app (cf. [8] for a detailed description and Fig. 5). The cellular radio transmits the recognized ADLs, EDLs to the LTAS server via http. In addition to communicating the wearer's geographic position by SMS the smartwatch app also establishes speech connections to the wearer of the smartwatch. The smartwatch app will also directly communicate with the stationary assistance subsystem, e.g. to inform it about the departure of the wearer from home and his return.
3. The **long-term analysis server LTAS**, which collects the ADLs, EDLs transmitted by 1, 2 and performs the long-term statistical analysis of the data. A smartwatch app accessing the server shows the last performed ADLs and vital signs of the person in need of support on demand. The LTAS will proactively inform relatives and/or authorized persons about substantial deviations of the usual lifecycle of the person in need of support. If configured, the server will

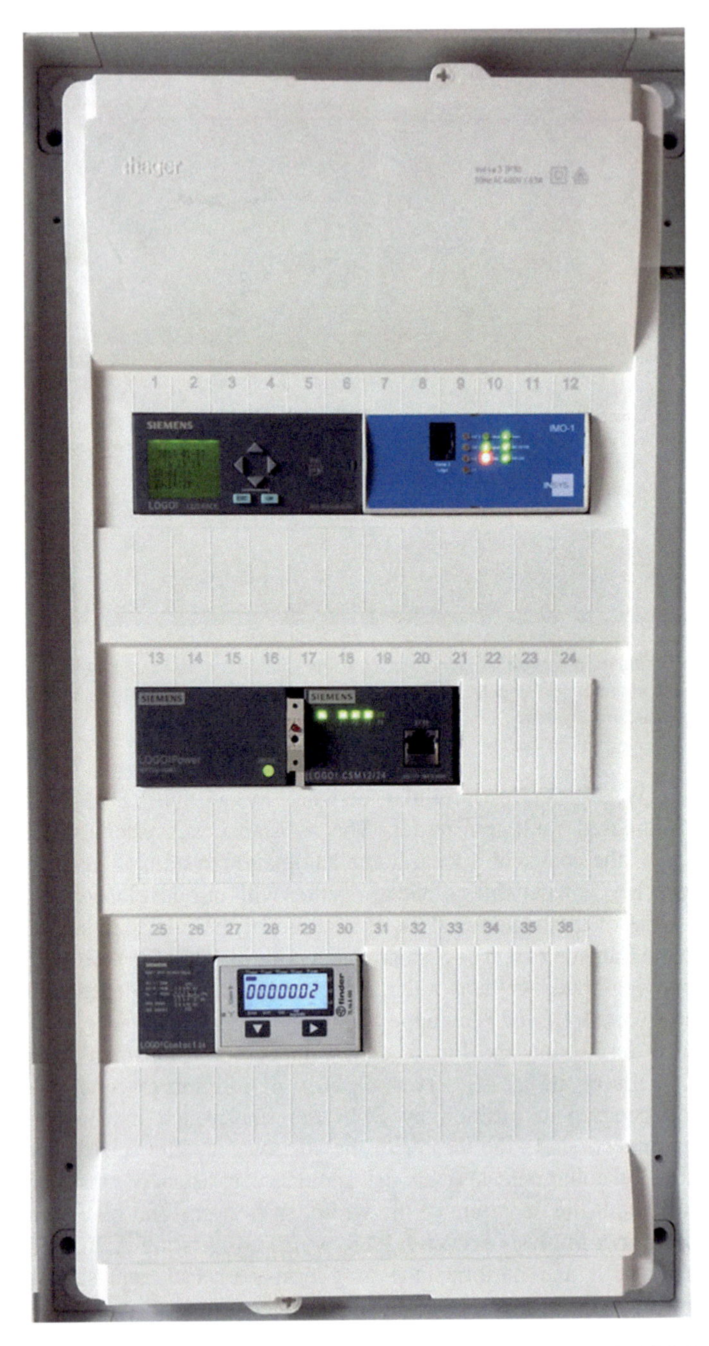

Fig. 3 Center of stationary assistance subsystem with Siemens LOGO™ SPS/PLC and insys IMO-1™ rule-based fault transmitter and GPRS router (*upper row*), power supply, industry IP switch (*middle row*) and electric power sensor and contactor for the electric stove (*lower row*) in a standard 3-row junction box

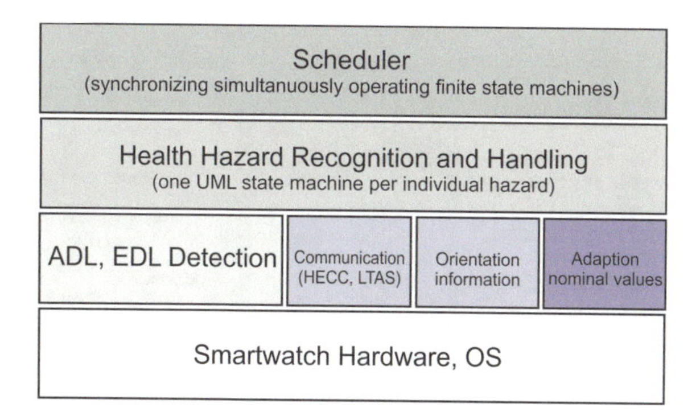

Fig. 4 Three layer smartwatch app architecture

Fig. 5 Smartwatch Samsung Gear™ S with assistance app displaying communication and orientation information (holidays, birthdays, ….) on the *left* and an advice to return to home after leaving an agreed area (geofencing) on the *right*

also provide a daily/weekly summary of all health-relevant activities of the monitored person (see Fig. 6).

Historically, the stationary subsystem (1) has started its productive use in the early spring of 2013, the smartwatch app (2) went into daily use in spring of 2015 and the LTAS (3) started its 24/7 operation on June 1, 2015 (Fig. 7).

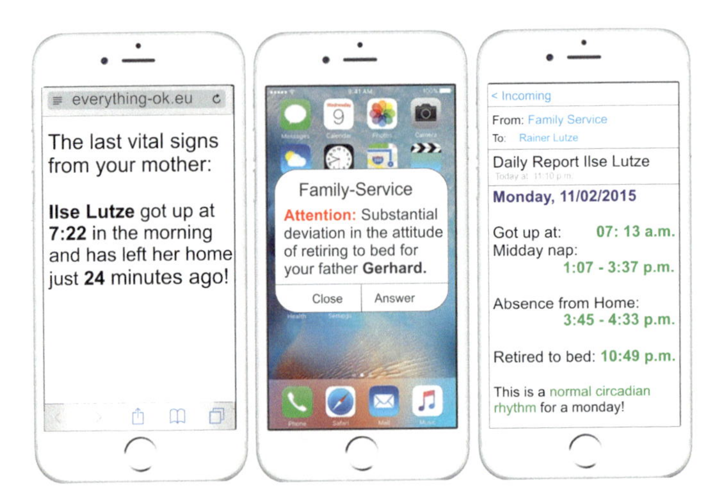

Fig. 6 Services for family members delivered by the LTAS: on demand requests for the last vital signs from the person in need of support/care (*left*), warning of substantial deviations of the used circadian cycle (*middle*), regular reports summarizing the last day (*right*)

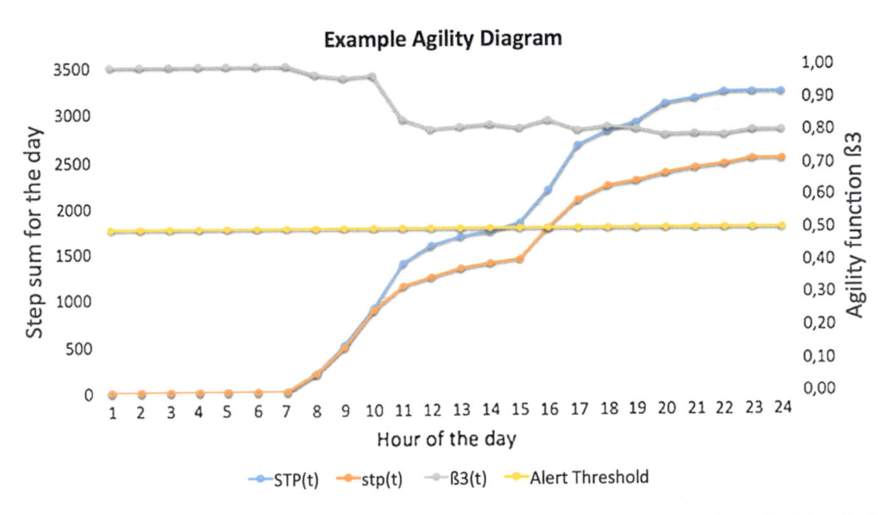

Fig. 7 Typical agility for a 24 h period based on an $\alpha = 0.1$ (giving heavy weights for historical values). As can been seen the agility value is far above the alert threshold indicating that no agility problems are present. Left scale denotes the accumulated steps (*orange* actual steps of the day, *blue* the estimated accumulated steps for the period, *grey* the agility value $ß_3$ and *yellow* the alert threshold for $ß_3$)

3 System Implementation

For the stationary assistance subsystem, many sensors and commercially available smart home systems have been evaluated and tested. For sensors, finally the decision in favor of high-end, two channel *presence sensors* was taken because standard movement sensors typically issued an unacceptable high rate of *false-negatives,* if people do not constantly move heavily (cf. [7] for details). For a discussion of potential alternative sensor technologies, see Sect. 5 of this document. One channel of the sensors is used for direct control of the ambient lighting; the second channel is used for sensing the presence of a person to the assistance subsystem.

The local usage of the sensors also for room lighting control was deliberately chosen to demonstrate the *comfort value* of the system to the residents on a daily basis. This automatic ambient lighting control was initially in fact judged by the inhabitants of the home as the most important and valuable feature of the new technology for them, although they were aware of the much more sophisticated assistance and monitoring technology in the background. The presence of this subsidiary monitoring technology for health hazards in the household was creating only awareness when health hazard pre-alert/alerts occurred from time to time.

Typically, the available sensor basis of commercially available (end user) smart home systems, (e.g. RWE Smart Home …), is not reliable enough compared to our presence sensors chosen. More important, it was not possible to implement the sensor fusion algorithms and finite state machines which are at the heart of the monitoring process (cf. [7, 9]) in a necessary fine-grained and precise way. Therefore we finally decided for an implementation based on the Siemens LOGO™ SPS/PLC and Insys IMO™ GPRS router/rule-based fault transmitter, which—as proven industrial components—work very reliable also in the presence of casual power brownouts and blackouts.

The knowledge for ADL, EDL recognition and the local handling of health hazards on the smartwatch is empirical, best practice knowledge which is growing and changing on a daily basis. The maintenance of the software encoding this knowl-edge with economic costs is a severe challenge, especially because all health hazards must be dealt with simultaneously and the hazard handling is intertwined in its execution.

We have developed a three layer architecture for smartwatch apps which allows to separate this knowledge into independent, small manageable chunks (cf. [10] and Figs. 4, 8):

- The *lower layer* contains the ADL and EDL detection based on sensor fusion. For simple EDLs, like *leaving resp. reentering an agreed vicinity around the home,* this detection will be done by trigonometric math calculations based on the current GPS sensor data. Complex ADL detection, for example for the *detection of fluid ingestion*, drinking (see [11–13]) utilizes neuronal networks or

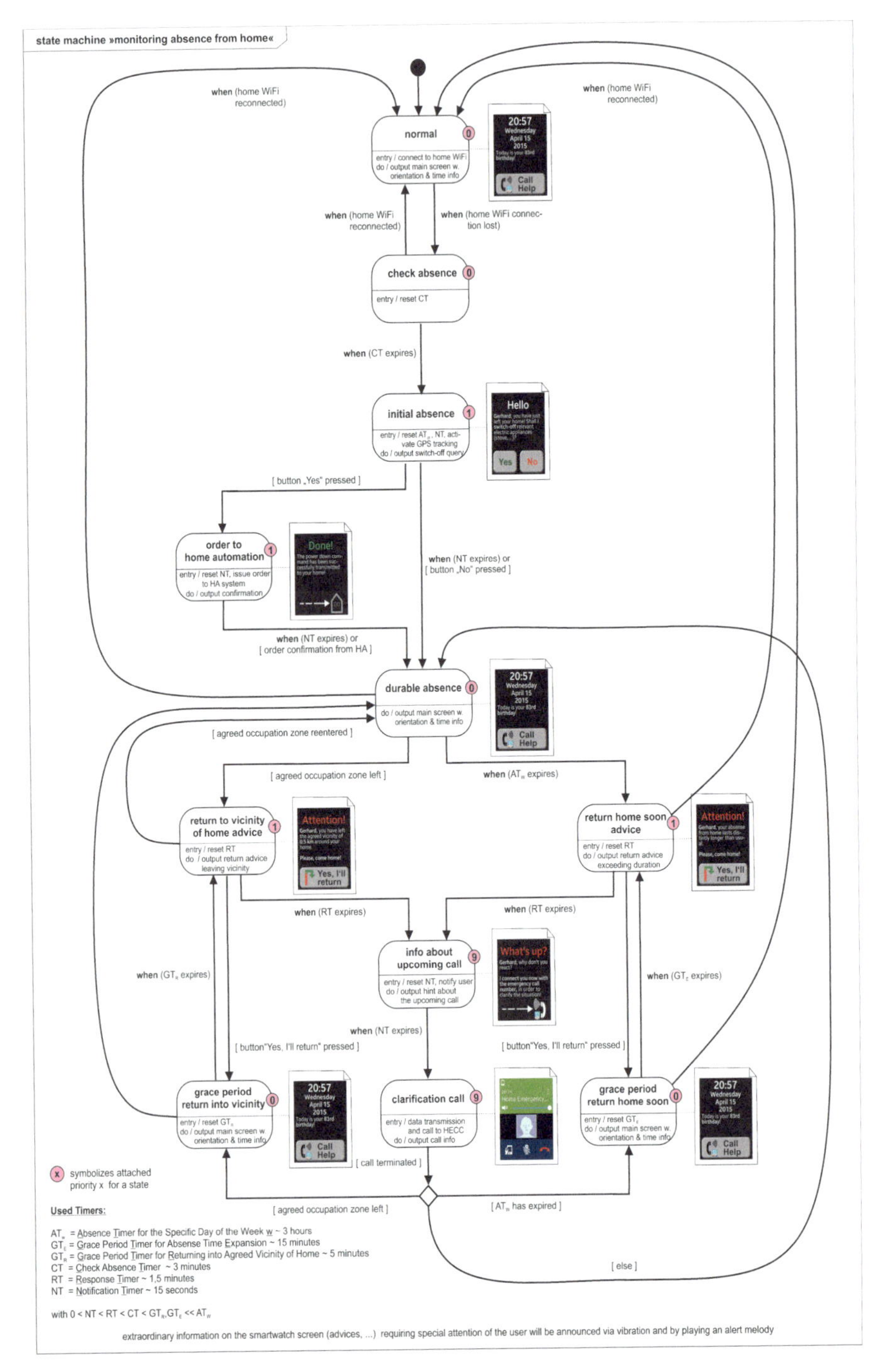

Fig. 8 Finite state machine for joint handling the health hazards resulting from the ADLs: *»absence from home«, »runaway situation«*

statistic regression methods for achieving the task. The nets resp. statistic parameters have to be trained before by several hundreds of supervised samples, in order to achieve the targeted precision and recall rate of a least 90% (cf. [13] for details of this process, which includes data conditioning and mining of the sample data on standard PCs). In addition this software layer contains the necessary functionality for managing speech connections, calls with the smartwatch and provides the necessary *orientation information* presented to the smartwatch wearer. Such orientation information—see Fig. 5—has been proven to be of high substantial value especially for persons with MCI or beginning dementia, in order to compensate the effect of their failing memory in communication situations with other persons.

- The *medium layer* on top of the ADL/EDL detection layer comprises finite state machines for recognition and handling of health hazards. A single state machine represents, in a declarative way, the processing of an *individual health hazard* by the smartwatch (cf. [10] for details). The finite state machine is described by its state transition table, including the corresponding actions to be executed when entering a state, and the state contents describing the output of the machine on the smartwatch screen while being in a specific state. For example, Fig. 8 describes the handling of an excessive absence from home either by (i) leaving a agreed vicinity around the home and/or (ii) extending the outside stay beyond an agreed maximum duration.
- The *upper layer* contains the central scheduler for synchronizing the simultaneous operations of the finite state machines for the individual hazards, thus allowing the smartwatch to monitor different health hazards simultaneously (e.g. *leaving agreed areas, excessive absence, falls, insufficient liquid ingestion, abnormal heart rates, …*). The scheduling algorithm selects the content to be displayed resp. interaction sequence with the wearer to be performed at a specific point in time, selecting the actual state with highest priority from all state machine (cf. [10] for details of the scheduling algorithm).

With the current generation of smartwatches, the high power consumption of the GPS sensor and the limited computational power of the CPU in continuously condensing all the sensor signals and comparing them against the trained patterns for ADL/EDL recognition limits the usage of the smartwatch to at most 18 h before the watch needs to be recharged. The smartwatch therefore is not a 24 h assistance device, but can only be used between rising up in the morning and retiring to bed at night. At night time the smartwatch device will be typically recharged. Assistance during this time can only be provided via the stationary assistance subsystem.

A direct communication link between the smartwatches and the stationary assistance subsystem assures that the departure from home and a later arrival can be managed without involving the long-term analysis server LTAS. The presence of the smartwatch wearer at home is detected via accessibility of the home Wi-Fi with a known SSID, thus allowing to detect the departure from home by loss of the Wi-Fi signal and later return by reconnecting to the home Wi-Fi. This is much more energy preserving than using GPS. The stationary assistance system—acting as a home automation system—can then switch-off critical electric loads, e.g. the electric stove, during the period of absence.

The LTAS records all EDLs/ADLs reported from the stationary assistance subsystems and the smartwatch in a relational database. Matching EDLs like departure from home and later arrival will be further condensed to a common ADL, e.g. the period of absence from home. In order to recognize substantial deviations from daily routine resp. the learned circadian cycle, the LTAS needs to be trained first for at least one week. During this week it will observe and record the nominal values for the presence, the duration of ADLs and the intervals between different ADLs on a weekday specific basis. Later on the nominal values will be adapted by a time series analysis with the actual values measured in subsequent weeks (see Sect. 3.3). Based on this information, the LTAS can then deliver three subtypes of services (cf. Fig. 6) to family members, relatives, caregivers or agents on duty in a home emergency call center during an incoming call from the home or from a smartwatch wearer.

3.1 EDLs and ADLs

Currently, the system recognizes and analyzes the following EDLs (event of daily living):

E_1 fall, tumbling
E_2 leaving home
E_3 returning to home
E_4 leaving a agreed vicinity around the home
E_5 returning into a agreed vicinity around the home
E_6 getting-up in the morning
E_7 retiring to bed at nighttime
E_8 falling asleep for a nap
E_9 awaking from a nap
E_{10} low battery situation of the smartwatch.

All EDLs have as their characteristic attribute the time t, at which they happen, but may have additional attributes (e.g. the actual battery level for E_{10}).

An ADL (activity of daily living) A_i is either atomic (e.g. A_5, A_6) or structured and then characterized by two EDLs e_1, e_2 happening at times t1, t2 with $t1 < t2$ as their start and end events, which we denote as $A_i[e_1, e_2]$ and thus A_i has a duration $ta = t2 - t1$. If for such a structured A_i, its characteristic start event e_1 has been recently recognized, but the end event e_2 is pending, A_i will be denominated as *ongoing*. See Fig. 9 for a depiction of the structured ADLs A_1 to A_4.

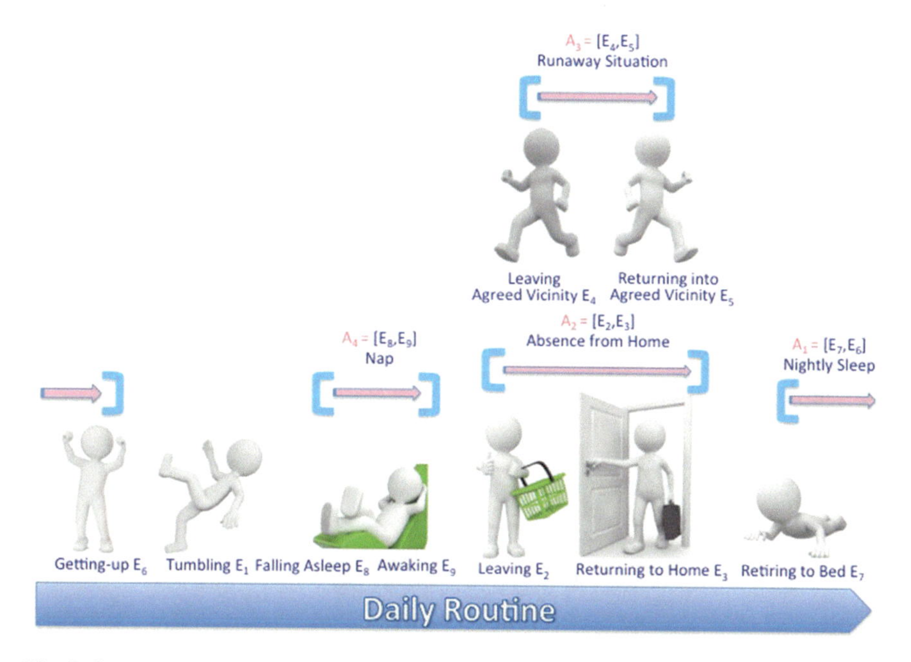

Fig. 9 Some EDLs and ADLs, including their specific starting and ending events within the daily routine

A_1 Bedtime, nightly sleep, defined by $A_1[E_7, E_6]$
A_2 Absence from home, defined by $A_2[E_2, E_3]$
A_3 Runaway situation, defined by $A_3[E_4, E_5]$
A_4 (Midday) nap, defined by $A_4[E_8, E_9]$
A_5 Visit to the toilet
A_6 Fluid ingestion, drinking.

Also ADLs have attributes, in general their actual duration ta, but also more specific attributes, as the time spend in the toilet room (excluding the way *to* and *from* the toilet) for A_5 or the amount for fluid ingested for $A_{6,}$ (the determination of the type of ingested fluid still an unsolved issue in our work).

3.2 Wellbeing Calculation and Monitoring

For the wellbeing function w we propose a (functional) combination of at least the three wellbeing aspects measuring the *inactivity* (β_1) of the person in need of support, *the excess duration of ongoing activities* (β_2) and the *agility* of the person in need of support (β_3). For these aspects we define three sub-functions β_1, β_2, β_3 mapping SensoricEvents $\rightarrow$ [0, 1], $[0, 1] \subset \mathfrak{R}$. A function value of 1 denotes ideal wellbeing. A health situation is judged the more dangerous the more those values decrease. If the w function value falls below a defined threshold, e.g. 0.5, an automatic alert will be issued. With respect to the following definitions this threshold value of 0.5 will be reached if the related *current* value for ADLs deviates by 70% from the specified *nominal* value.

When no recognized ADL in the household is taking place, $ß_1$, the wellbeing sub-function for **inactivity** measurement, is applied based on the definition in [3], by:

$$e^{-t/2*T}$$

where t is the current (time) duration of inactivity since completion of the last ADL, and T is the specific average inactivity between ADLs learned from the past for the current day of the week.[2]

On the opposite, as long as a recognized ADL is ongoing, $ß_2$, the wellbeing sub-function for the measurement of **excess duration** of this specific ADLs will be applied, which has been defined in [3] to:

$$\begin{array}{ll} e^{(TN-ta)/TN}, & \text{for } ta > TN \\ 1, & \text{otherwise} \end{array}$$

where ta is the actual duration of the (ongoing) ADL and TN is the specific maximum duration of the corresponding recognized ADL in a *normal situation* learned from the past for the current day of the week (Assuming at least a typical one week cycle for calculating the specific, possibly varying T, TN, SN, F nominal values for the individual days of a week. If the current day under consideration is a national or local holiday, based on the Western cultural context, the values for the last Sunday will be used instead.). For simplicity reasons of calculating $ß_2$, it is not always necessary to know which ADL exactly is happening, but the set of all ADLs will be partitioned into categories with respect to similar typical execution times, and then all ADLs within the respective partition will have the same maximum execution time TN. $ß_2$ will then be calculated with the specific TN value based on the ***category*** of the recognized ADL.

Independent from detected EDLs, ADLs occurring resp. being carried, we propose for the new wellbeing sub-function $ß_3$ determining the **agility** of the elderly subject:

$$\begin{array}{ll} e^{(stp(t)-STP(t))/STP(t)}, & \text{for } stp(t) < STP(t) \text{ and } NOT(E_1) \\ 1, & \text{for } stp(t) \geq STP(t) \text{ and } NOT(E_1) \\ 0, & \text{if event } E_1 \text{ has been detected} \end{array}$$

[2]Assuming at least a typical one week cycle for calculating the specific, possibly varying T, TN, SN, F nominal values for the individual days of a week. If the current day under consideration is a national or local holiday, based on the Western cultural context, the values for the last Sunday will be used instead.

where stp(t) is the sum of steps performed during the current day until actual time t, F(t) is the cumulative distribution function of steps over the day, SN is the specific total number of steps learned from the past for the current day of the week and with STP(t) = SN * F(t) estimated from the nominal step sum for the current day at time t. $ß_3$ will be calculated all over the day. An advantage of $ß_3$ is that it does not rely on ADL detection and thus counterbalances the up-to-date dependency of the well-being calculation from the plenitude of recognized ADLs. The sub function $ß_3$ will depend primarily on the (counted) steps measured by the smartwatch over the course of the day and taking into account the cumulative distribution of those steps until time t of the specific day learned from the past. A recognized fall will immediately effect an alert (see below).

Finally, the wellbeing function w: SensoricEvents → [0, 1], $[0, 1] \subset \mathfrak{R}$, will be formally defined as:

$$w = \min\{ß_1, ß_2, ß_3\}$$

This means that whenever the *inactivity* (missing any recognized ADL) or the *excess duration* of an ongoing activity category or the lacking *agility* gets critical and the w value falls below 0.5, a *health hazard alert* will be issued.

3.3 Learning the Nominal Values

For the acquisition of the nominal values for T, the TNs for each ADL category, SN and the cumulative movement distribution function F(t) a seasonal cycle of one week and a preceding initial training phase of a week will be used. For each day $s \in$ Nat, s = 1, 2… 7 in the training phase and the following weeks, the (weekday) specific T_s, the TN_s for each ADL category, SN_s and $F_s(t)$ values resp. distributions are estimated:

- **Inactivity**: For all ADLs A_1 … A_X detected on this day s, for T_s the average $(I_1 + I_2 \ldots + I_{x-1})/X-1$ of all inactivity periods $I_1 \ldots I_{x-1}$ will be used, where I_n denotes the inactivity period between A_n and A_{n+1}.
- **Excess duration**: For all activity (time) durations $d_1, \ldots d_X$ of those ADL instances $A_1, \ldots A_X$ detected on this day s for a specific ADL category, for TN_s for this ADL category the maximum duration $MAX_{i=1,\ldots,x}(d_i)$ will be used.
- **Agility**: For each hour h = 0… 23 of the day, the sum of all counted steps achieved by the end of a hour (from the beginning of the day) will be tabulated for computing the initial $F_s(h)$ distribution. SN_s will be the total number of all steps counted for this day s.

For projecting these values for the next time for day s, s > 7, we use the exponential moving average already proposed in [14] for MA = T, TN, SN, F:

$$MA_s = \alpha^* O_{s-7} + (1-\alpha)^* (MA_{s-7} + TR_{s-1})$$

where O_x is the observed value for day x (computed by the method described above) and MA_x is the computed moving average for day x and α, $0 \leq \alpha \leq 1$, $\alpha \in \mathfrak{R}$, is a “smoothing constant”, which may give more relative weight either to observed values in the week before or the predicted value, moving average, for the specific day in the week before. Initially $MA_i = O_i$ for all i = 1, …, 7. The seasonality factor TR_x covers a potential *data trend* by accounting the differences in O_x values for subsequent days of the last week (cf. [3]) and will be contributing starting from the third week (s = 15), for all prior days i $\leq$ 14: $TR_i = 0$. TR_x is defined by:

$$1/7^* ((Ox - O_{x-7})/7 + (O_{x-1} - O_{x-8})/7 + \ldots + (O_{x-6} - O_{x-13})/7)$$

Figure 7 shows a typical example of an elderly for the agility parameter:

4 Experiences

4.1 Development of Functionality Over the Years

During the initial years of operation, when the stationary assistance subsystem was the only available component, the main use of the assistance subsystem was: (i) the *monitoring of inactivity periods* via $ß_1$ at day time (*“inactivity analysis”*, cf. [6, 15]) and (ii) the $ß_2$ monitoring of an *excessive duration of the* ADLs *»bedtime«and» visit to the toilet«*. At that time, we used an additional ADL *»motion at home«* in order to have a more structured daytime. Although, in a multi-person household—the person in need of support and his spouse, guests in the home—this ADL did not really allowed to conclude specifically about the health state of the person in need of support. The introduction of the smartwatch therefore brought a substantial progress, because the movement information from now on could be assigned directly to the person in need of support and supervised by $ß_3$; thus the recognition of this initial ADL *»motion at home«* was abandoned with the availability of the new smartwatches. For the recognition of ADLs A_1 *»bedtime«*, A_5 *»visit to the toilet«*, the stationary assistance subsystem achieved a durable precision rate of about 98% based on the false positive and negative alert elimination techniques described in [7]. In essence, the utilized finite state machine did so in heuristically applying the fact that persons cannot arbitrarily appear or disappear in the specific rooms of the home other than enabled by (i) the connectivity of the rooms and (ii) based on their room specific location inside the home (or outside) up to now.

With the finite state machines in the background, especially the sequencing of activities for a nightly visit of the toilet, ADL A_5, starting and ending from and in the bedroom was supervised. The transit times from the bedroom to the toilet and back again were so characteristic for the individual persons that these toilet visits could be clearly assigned to the individual persons. We used this information to automatically fill-up the daily nursing record requested by the health insurance. We further observed in the course of the progressing dementia illness of the person in need of support that the transit time necessary to pass a distance of about 10 m grew within the operation time of the assistance system nearly by factor of ten due to the proliferating disorientation of the patient. Pre-alerts of the assistance system at that time primarily addressed resp. targeted to give attention to resp. to wake-up the care-giving spouse or family member present in the home in order to verify that everything was ok with the person in need of support. With the progressing dementia and the noticeable exhaustion of the care-giving persons by the more and more demanding care-giving task, the nightly $ß_2$ induced pre-alerts gained more and more practical importance over the course of the years. The exhaustion of the caregivers was not only caused by number of nightly toilet visits continuously increasing during the progress of the dementia illness and accompanied by an increasing frequency of $ß_2$ excess duration alerts for those visits. Also the increasing frequency of nightly unrest periods in the bedroom, where the person in need of support was not able to sleep continuously more than 3 h, contributed to this. Similar scenarios have been reported to us by many care-giving relatives with respect to their family members in need of support and impaired by dementia. At this stage of the dementia illness, two years after the introduction of the assistance technology, the real value of the installed technology in providing much more support than automatic ambient lighting was fully understood and perceived by the care-giving spouse and family relatives.

If a local pre-alert remains unanswered, the stationary assistance subsystem places an automatic alert via SMS and E-Mail. The following external handling of such alerts suffers from the same limitations as for the traditional home emergency call devices: if the check-back telephone call to the home also will not be answered, complicated research has to be initiated by the alert serving agent. This includes a potential questionnaire of local neighbors by telephone …, in order to decide whether a costly on-site emergency intervention shall be done rsp. an intervention team shall be send out to the alerting home.

Now, with the availability of the LTAS, the alert serving party can immediately retrieve contextual information about the recent history of ADLs in the corresponding home, even if the check-back call will not be answered. This allows faster and more fact based decisions about sending out the emergency intervention team, especially at nighttime, when no neighbors can be asked. Moreover, with the cellular telephone included in the smartwatch worn at the wrist of the person in need of support, the likelihood of establishing a successful speech connection between the alert serving agent and the person in need of support is substantially increased. If a health hazard has been detected by the smartwatch and a local pre-alert has not been answered by the wearer of the watch (terminating false alerts) the smartwatch app initiates the telephone call. For such an automatically established direct speech connection from the person in need of support, the presence of a severe health hazard typically has to be assumed, if the person is not able to respond verbally anymore.

For the computation of the wellbeing function w, the computation of its constituents and—partially—the adaption of the nominal values has been located to the stationary subsystem (for $ß_1$, $ß_2$) and the smartwatch (for $ß_2$, $ß_3$), in order to guarantee the functionality and the issuing of corresponding pre-alerts/alerts in a most reliable way, even if no other system component is momentarily at hand. The notion of a finite state machine has been proven to be a suitable implementation of the abstract wellbeing definitions above, for example see Fig. 8 (and cf. [10]) for a local implementation for the $ß_2$ joint health hazard handling of the ADLs A_2, A_3 on the smartwatch.

Overall the smartwatches were welcomed and have been accepted by the elderly people of our case study as non-stigmatizing, multipurpose devices with a readable, large enough display. In contrast to wireless remote units of the classic home emergency call devices, which were categorically rejected by the elderly because they felt (with a real age of 85 and more) "too young" in order to use these devices! Especially the orientation information provided by the app (see Fig. 5) was valued as the feature with highest value for the person in need of support because it enables a continued participation in social interaction without directly disclosing the mental disability; an unobtrusive glance at the watch compensates—or at least camouflages—the failing personal memories.

From the assistance system perspective the smartwatch app complements the functionality of the system in that EDLs like » falls « can be directly recognized which so far could only be indirectly observed via its potential consequences by the stationary assistance subsystem. In addition, the smartwatch app enables the detection of more fine-grained ADLs like *»midday naps«* or *»liquid ingestion«* , ADLs, for which it was so far impossible to observe them via the presence sensors of the stationary subsystem. And for outdoor stays, the direct intertwining of the stationary subsystem and the smartwatch allows the fully automatic recognition of the ADL *»absence from home«* . For the stationary subsystem alone, so far the subsystem had to be informed manually at least about the beginning of an outdoor stay in order to prevent false alarms. In practice, this often was forgotten and consequently caused such irritating false alarms. And, last but not least, the smartwatch app enables the reach of the assistance beyond the spatial borders of the home and includes protection for strolling, shopping, visits to the doctor. Especially a *»runaway situation«* , which not infrequently results in a deathly outcome and produces extensive and costly public search actions, can be dealt with in very short-term. On the other hand the current smartwatch app is of productive use only between getting up in the morning and retiring to bed at nighttime, while the stationary assistance is on duty at nighttime when the smartwatch needs to be recharged.

From the perspective of the care-giving persons and distant family member, the main advantage of the smartwatch app is the permanent reachability of the persons in need of support due to the integrated cellular radio of the smartwatch. Especially for elderly people of high age the general tendency is not to carry their mobile phones with them when they leave their home because they have not been acquainted to this during their working life. With the smartwatch typically continuously worn during the whole day, a forgotten or deliberately not carried along smartphone has no consequences with respect to reachability and the functionality of the assistance services.

4.2 Usability

The stationary assistance subsystem and the smartwatch differ substantially in their usability. As a subsidiary system, the stationary assistance subsystem will not *actively* operated by the person in need of support. At most, the person in need of support has to *react* to prealerts in order to signalize that everything is currently ok by cancelling those prealerts. The person in need of support needs not to be aware of the assistance technology "in the background" rsp. is not required to have any in-depth understanding of the stationary assistance subsystem. This makes the technology especially suited also for people with mental disabilities, e.g. caused by MCI or dementia.

In contrast, use of the smartwatch requires a basic *"digital competence"*, in order to gain a beneficial use by the wearer. This starts with insight in the necessity of recharging the electronic device each night (as it is self-evident practice for most smartphones). It includes basic capabilities of operating the GUI of the smartwatch and circumventing the use of the default installed, but partially complex, manufacturer provided system apps on the watch. Also, the wearer needs to have a basic understanding about the many assistance functions of the app, which only can deliver effective support, if he/she are really wearing the device. Otherwise, the additional value of wearing the smartwatch in contrast to the used smartphone will not be perceived. And, for example, the wearer should have a basic understanding about the operation principles of GPS, by which the app can only track his/her exact position after a few moments with free sight to the open sky. It turned out also for the field test,[3] which followed our case study, that the prior knowledge or sharing of this *"digital competence"* was a mandatory precondition and decisive factor for the successful use of smartwatches. This recommends the application of smartwatches for elderly persons primarily not of very high age, which ideally already have a mobile phone usage background from their prior active work life.

[3]The field is currently carried out with a group of persons in need of support in a rural area of the State of Hesse, Germany, near Marburg, and the urban Rhein-Main area, in cooperation with the Deutsche Rotes Kreuz (DRK) as the caritative service provider. The field test is managed and scientifically accompanied/ assessed by the Frankfurt University or Applied Science (UAS) in the scope of the Hessian state excellence program "LOEWE"'s GSMTS project [16, 17]).

4.3 Practicability

Another aspect concerns the charging procedure, which in our case of the Samsung Gear™ S requires a mechanical precision attachment of the watch to a charging adapter. This requires fine motoric capabilities typically not present anymore for elderly people and caused regular handling problems during our study. We therefore included a specific SMS to the caregivers and also an alert to the LTAS requesting recharging in case of an ongoing low battery situation (event E_{10})! Fortunately, the manufacturers have addressed this problem meanwhile. An inductive charging without mechanical contacting of the watch—as proposed by the Moto 360™ and Apple Watch™—will be the future general standard (now also featured by the Samsung successor model Gear™ S2).

4.4 Synergies by Integrating Stationary and Wearable Information in the LFAS

The integration of delivered information from the stationary subsystem and the smartwatches in the LFAS has been used especially for mutually completing the information from both sources and thus reducing false alerts. For example, if the smartwatch wearer has left his home and the smartwatch battery would fade out during his absence, no return information will be later generated by the smartwatch. In such a situation, if the stationary presence sensor signals the presence of a person at home at a time conforming with the usual return time of the smartwatch wearer for this day of the week, it will be assumed by the LFAS that this present person in the household will be in fact the smartwatch wearer and no $ß_2$ excessive absence alert for A_2 will be communicated. In the opposite case, if no one returns to home in due time, even in case of an empty battery (and therefore a silent smartwatch), the LFAS will communicate an excessive $ß_2$ absence alert based on the prior E_2 information about leaving the home. Also E_1 »*fall*« EDLs occurring in the household and eventually not detected by a worn smartwatch, will be finally detected by the presence sensors due to overtime presence at the location of the fall, if the person in need of support should not move resp. leave the location after the fall.

5 Discussion

The choice of the most suitable sensor technology for an assistance system is an ongoing discussion based on the respective technologic progress. In addition to the stigmatization and cost/value benefits aspects mentioned above, additional aspects like the estimated maturity of a sensor technology and its foreseeable support and maintenance costs as well as the subjective fear of (especially German) users of "electro smog" had to be additionally taken into account.

In [6, 18] comprehensive overviews of stationary, *wearables* and *image-based sensor* projects *are* presented. The up-to-date overview in [18] has been elaborated in the scope of the SPHERE project in order to explore the strength and limitations of the different sensor technologies for monitoring various health conditions. [19] focuses on the usage of such sensors especially for the purpose of fall detection. In our estimation, the reported slightly decreased recognition rate of falls by wearables, smartwatch sensors is (more than) counterbalanced by their universality of use also outside the home, the substantially reduced price of wearables in contrast to floor sensors for fall detection and the non existing privacy concerns and installation problems at the room ceiling in contrast to any kind of image based fall sensors.

In [20], an 80% recognition rate of 19 fine-grained ADLs is reported, if active sensors are not only placed on the wrist, but additionally on the legs, waist and back of persons, and active Bluetooth beacons had been placed in each room of the home. In our estimation and from a practical perspective, the unquestionable achievements in ADL recognition by this plenitude of sensors has to be questioned against the additional 2, 4 GHz radiation emitted by the beacons and the burden of attaching and wearing so many sensors on a daily basis. Our approach builds on the opposite principle: identifying only wellbeing and health critical ADLs with a minimal number of sensors, but with a maximum precision. The non-intrusive (electric) appliance load monitoring of [21, 22], which requires only one central digital load monitor at the electricity meters in the central junction box of the household would much more comply with our spirit.[4]

In [23] an interesting approach focusing on *things* in contrast to our *movement of persons* centered approach for ADL detection is presented. In the thing based approach all relevant things (dishwasher and washing-machine safe) in the household are equipped with low cost RFID tags which can be attached to most tableware. In this approach, the ADLs will be concluded from the specific movements of things caused by human activity. The RFID tags attached to the things will be periodically powered by multiple antennas covered in the walls of the home. These antennas also receive the response from the RFID tags, including their individual ids, in a very short time frame after inductively powering the tags. From the response signals received by different antennas, the position of the things will be monitored by trilateration. The sequence of moving positions will then be used for activity recognition via the help of Bayesian networks. Although this technology is limited to the home by its nature, its strength is a fine-grained analysis of eating behavior, one of the weak aspects of our approach.

[4]We finally decided against this sensor technology due to its susceptibility against smart home functions like "virtual inhabitant" or "presence simulation", which deliberately produce confusing load patterns in the case of absence of the inhabitants.

In [24, 25] a monitoring approach based on a wireless sensor network (WSN) for the home is presented, which utilizes specific pressure resp. occupancy sensors for chairs and beds in addition to standard PIR movements sensors. Our view is that the costs for such dedicated sensor devices, including initial installation in the furniture and regular maintenance for providing electric power to the sensors, will not be justified for the future by the gain in the precision of ADL recognition. With our high-quality presence sensors and in combination with the smartwatch sensors and the finite state machines in the background, occupancy in bed and sitting in a room can be typically recognized resp. concluded from these more universal devices.

6 Conclusions and Future Work

The amendment of the original stationary assistance subsystems by smartwatches and the LTAS indeed increases the perceived value of the assistance technology:

- for the users, by providing more mobile services, e.g. the wearable orientation information in Fig. 5, and extending the reach of support services beyond the spatial boundaries of the home,
- for the system, by recognizing more ADLs (e.g. » runaway situation « , » midday nap « , » liquid ingestion «), detecting EDLs directly rsp. in a shorter time frame (e.g. » fall «) or recognizing ADLs now in a more reliable and fully automatic way (e.g. » absence from home «) and by integrating wearable and stationary sensor information aiming at a reduced number of false alerts,
- for parties in charge of reacting to external alerts (distant family members, home emergency call center agents), by providing contextual information for more fact based decisions about what happened to the person in need of support at the distant home or on the go (e.g. services in Fig. 6).

Our future work will be directed towards the elaboration of the developed data mining based activity recognition technology ([11–13]) for improving the resource efficient and stable execution on the smartwatch, for a more reliable detection of falls and for detecting more fine grained ADLs on the basis of this technology like *»teeth-brushing«, »hand washing« and »combing«* as symptoms of a well-managed life. Another focal point will be the enforced utilization of the LTAS for identifying and filtering—in result: reducing—potential false alerts by heuristically combining all relevant information from wearables and stationary sensors. This combination of historical and current information about EDLs, ADLs will also be used by the LTAS for producing fact-based, trustworthy and comprehensible natural language justifications for communicated alerts.

Also hardware progress on the sensor equipment of the smartwatches, for example for *pulse oximetry* carried out within the smartwatch on the wrist, would be highly welcome. In the long-term, we do foresee the evolution of the smartwatch app towards a regulated medical software product, at least, when together with today's *heart rate measurement* and the pulse-oximetry in the future the *blood pressure* could be also measured with the smartwatch resp. its wrist band. The basic research for this, namely to utilize the transition time of the arterial pressure valve from the left edge of the wrist band to its right edge, a time period which is strongly related to the individual blood pressure, is already on the way (ETH Zürich, cf. [26]). With (i) current heart rate, (ii) the arterial oxygen saturation of the blood and iii) the blood pressure, the most essential three medical parameters for a diagnosis of vital health hazards would be available for the smartwatch app. This regulated medical software product would run on a smartwatch consumer hardware and thus would be required to cope with potentially wrong and/or inconsistent hardware sensor values of the consumer hardware without leading to a wrong diagnosis (cf. [27] for those requirements).

This will elevate the added value of the smartwatch app and level of possible assistance by the app to a complete new quality.

References

1. Künemund, H.: Chancen und Herausforderungen assistiver Technik: Nutzerbedarfe und Technikakzeptanz im Alter. Technikfolgenabschät-tzung – Theorie und Praxis **24**(2), 28–35 (2015) (in Ger-man)
2. Juchli, L.: Krankenpflege - Praxis und Theorie der Gesundheitsförderung und Pflege Kranker. Georg Thieme Verlag Stuttgart-New York (1987)
3. Suryadevara, N.K., Mukhopadhyvay, S.C.: Determining wellness through an ambient assisted living environment. IEEE Intell. Syst. **29**(3), 30–37 (2014)
4. Fachinger, U., Koch, H., Henke, K.-D., Troppens, S., Braeseke, G., Merda, M.: Ökonomische Potenziale altersgerechter Assistenzsysteme: Er-gebnisse der Studie zu Ökonomischen Potenzialen und neuartigen Geschäftsmodellen im Bereich Altersgerechte Assistenzsysteme. (2012) (in German)
5. Kirkpatrick, K.: Sensors for seniors. ACM Commun. **57**(12), 17–19 (2014)
6. Lutze, R.: Assistance systems at home—economics, acceptance barriers and multivalent utility. In: Proceedings of 4 German AAL Conference, 25./26.01.2011 Berlin/Germany, VDE-Press, Paper 7.1
7. Lutze, R.: Road Capability Aspects of Assistance Systems at Home. In: Proceedings of 7 German AAL Conference, 21./22.01.2014 Berlin/Germany, VDE Press, Paper S22.2 (in German)
8. Lutze, R., Waldhör, K.: Smartwatches as next generation home emergency call systems. In: Proceedings of 8. German AAL Conference, 29./30.4.2015 Frankfurt/Germany, Paper A1, VDE Press (in German)
9. Danacher/J.-J.Lesage, M., Litz, L.: Indoor location tracking based on a discrete event model. In: Proceedings of 10th Int. Conference on Smart Homes and Health Telematics (ICOST 2012), LNCS 7251, pp. 262–265. Springer (2012)
10. Lutze, R., Waldhör, K.: A smartwatch software architecture for health hazard handling for elderly people. In: IEEE International Conference on HealthCare Informatics (ICHI), 21.–23.10.2015, Dallas/USA, pp. 356–361 (2015)
11. Waldhör, K., Lutze, R.: Effective support of care-giving relatives by smartwatches. In: Proceedings of 8. European Nursing Informatics Conference (ENI), 29./30.9.2015, Hall/Austria, (in German)

12. Baldauf, R., Lutze, R., Waldhör, K.: Dehydration prevention and effective support of elderly by the use of smartwatches. In: Proceedings of IEEE HealthCom Conference, 14–17.10.2015, pp. 404–409. Boston/USA (2015)
13. Waldhör, K., Baldauf, R.: Recognizing trinking ADLs in real time using smartwatches and data mining. In: Proceedings of Rapid Miner Wisdom/ Europe Conference, 31.8.2015 Ljubljana/Slovenia (2015, in press)
14. Suryadevara N.K., Mukhopadhyvay, S.C., Wang, R., Rayudu, R.K., Huang, Y.M.: Reliable measurement of wireless sensor network data for forecasting wellness of elderly at smart home. In: Proceedings of IEEE International Conference on Instrumentation and Measurement Technology (I2MTC), Minneapolis/USA, pp. 16–21 (2013)
15. Rodner, T., Litz, L.: Inactivity monitoring and automatic alerting—from theory into practice. In: Proceedings of 7. German AAL Conference, 21./22.01.2014 Berlin/Germany, VDE Press, Paper A17 (2014) (in German)
16. Rossberg, H., Reutzel, S., Klein, B.: Techniktransfer und Barriere-freiheit von assistiven Technologien – zur Verknüpfung von Hausnotruf und Wearables. In: Proceedings of 8. German AAL Conference, 29./30.4.2015 Frankfurt/Germany, Poster, VDE Press (2015) (in German)
17. Klein, B., Reutzel, S., Roßberg, H., Bienhaus, D., Lutze, R., Hofmann, J., Dallwitz, D., Sütö, S., Heinrich, H., Donath, M.: *Anbindung von Hausautomation und Wearables: Akzeptanz und Gebrauchstauglichkeit.* Zukunft Lebensräume Kongress; 20–21.04.2016; Frankfurt am Main, VDE-Verlag, pp. 109–115 (2016)
18. Zhu et.al.: Bridging e-health and the internet of things: the SPHERE project. IEEE Intell. Syst. **30**(4), 39–46 (2015)
19. Igual, R., Medrano, C., Plaza, I.: Challenges, issues and trends in fall detection systems. Bio Med. Eng. Online **12**(1), 66 (2013)
20. De, D., Bharti, P., Das, S.D., Chellappan, S.: Multimodal wearable sensing for fine-grained activity recognition in healthcare. IEEE Internet Comput. **19**(5), 26–35 (2015)
21. Wilken, O., Kramer, O., Steen, E., Hein, A.: Activity recognition using non-intrusive appliance load monitoring. In: PECCS 2014—Proceedings of the 4th International Conference on Pervasive and Embedded Computing and Communication Systems. SCITEPRESS, pp. 40–48
22. Clement, J., Ploennings, J., Kabitzsch, R.: Detecting activitities of daily living with smart meters. In:Wichert R., Klausing K. (eds.) Ambient Assisted Living 6 AAL Kongress 2013. Berlin, Germany, January 22–23, 2013, VDE / Springer 2014, pp. 143–160
23. Fortin-Simard, D., Bilodeau, J.S., Bouchard, K., Gaboury, S., Bouchard, B., Bouzouane, A.: Exploiting passive RFID technology for activity recognition in smart homes. IEEE Intell. Syst. **30**(4), 7–15 (2015)
24. Fernandez-Luque, J.F., Martinez, F.L., Zapata, G.D.J., Ruiz, F.: Ambient-assisted living system with capacitive occupancy sensor. Expert Syst. **31**(4) 378–388. Wiley Publishing 2013 (2014)
25. Suryadevara, N.K., Mukhopadhyvay, S.C.: Wireless sensor network based home monitoring system for wellness determination of elderly. IEEE Sens. J. **12**, 1965–1972 (2012)
26. Press release of the Swiss EMPA (Eidgenössische Materialprüfungs-anstalt) from June 12, 2013: Wristband revolutionises blood pressure measurement. www.empa.ch
27. Heidenreich, G., Neumann, G.: Software für Medizingeräte. Publicis Publishing, SIEMENS AG, Berlin und München (2015)

QuoVadis—Definition of Requirements and Conception for Interconnected Living in a Quarter for Dementia Patients

Alexander Gerka, Nadine Abmeier, Marie-Luise Schwarz, Stefanie Brinkmann-Gerdes, Marco Eichelberg and Andreas Hein

Abstract The choice of a suitable living and care arrangement for dementia patients is a difficult decision for the patients themselves and their relatives. This decision becomes even more difficult if only classical arrangements are considered. Such arrangements, like care in long term care facilities or care in the own apartment administered by the relatives, often do not address the needs of dementia patients appropriately or cause significant stress for the relatives. Therefore, a new living arrangement concept is developed in the research project QuoVadis, which is called "interconnected living in the quarter". In this new living arrangement dementia patients can stay in their own accommodation while they are integrated into a care concept at the same time. In a quarter, a "quarter manager" is responsible for organizing care and help for the dementia patients. To empower the quarter manager to support the residents, an intelligent sensor-based system is developed. This system will be integrated into the apartments of the dementia patients. This article presents the requirements definition for the technical system. Furthermore, a concept to fulfil these requirements is presented. As part of this concept, an adaptive decision support system (ADSS) will be presented.

A. Gerka (✉) · M. Eichelberg
OFFIS - Institute for Information Technology, Oldenburg, Germany
e-mail: gerka@offis.de

M. Eichelberg
e-mail: eichelberg@offis.de

N. Abmeier · M.-L. Schwarz
Johanniter-Unfall-Hilfe e.V, Berne, Germany
e-mail: Nadine.Abmeier@johanniter.de

M.-L. Schwarz
e-mail: Marie-Luise.Schwarz@johanniter.de

S. Brinkmann-Gerdes
GSG OLDENBURG Bau- und Wohngesellschaft mbH, Oldenburg, Germany
e-mail: stefaniebrinkmann-gerdes@gsg-oldenburg.de

A. Hein
Carl von Ossietzky University, Oldenburg, Germany
e-mail: Andreas.Hein@informatik.uni-oldenburg.de

R. Wichert and B. Mand (eds.), *Ambient Assisted Living*,
Advanced Technologies and Societal Change,
DOI 10.1007/978-3-319-52322-4_2

1 Introduction

Every person with dementia and his or her family face the same question: Which living arrangement is the most suitable? In many cases, elderly people prefer to stay at home rather than moving to long-term care facilities [1, 2]. The transition to care facilities has several disadvantages for dementia patients. One disadvantage is the necessity to adjust to new circumstances. Especially, dementia patients may have problems to adjust to their new environment [3]. Additionally, living in care facilities seems to have a negative impact on the self-perceived quality of life [4] as well as the mortality rate of dementia patients [5]. However, if people who suffer from dementia decide to stay in their own apartments, the situation can be very challenging for the patients and their relatives alike [6]. What this all amounts to is that people with dementia face risks and challenges, indifferent whether they stay at home or move to a long-term care facility home. To enable dementia patients to feel homey, while ensuring their safety and adequate accommodation conditions at the same time, novel living arrangements have to be developed. One of these alternative living arrangements is the assisted living community for dementia patients. In this living arrangement people with dementia share a household. In such an assisted living community there is always a carer or a caretaker available to provide support for the inhabitants if needed. Additionally, caretakers organize the activities of daily living. As a result, the residents are able to lead a self-determined life, while being cared for appropriately [7]. However, assisted living communities still require the dementia patients to move out of their own home.

The aim of the research project QuoVadis is to develop a new concept of "community living" that does not require dementia patients to move out of their well-known surroundings. In this concept, elements of assisted living communities will be transformed into a living quarter. Therefore, a local infrastructure is implemented in a quarter of the town or the village where dementia patients live. In each quarter a "quarter manager" is situated. The task of the quarter manager is to support the dementia patients and their relatives. In order to empower the quarter manager to support the patients without visiting them personally every time, ambient assistance technologies will be used in the apartments of the dementia patients. An overview of the relevant technologies is presented in this chapter. Furthermore, the requirements of patients in a quarter-based care concept (Chap. 3) and the concept to meet these requirements (Chap. 4) are presented in this paper.

The developed concept will be tested in two quarters in Oldenburg and Brake, Germany. Each quarter will comprise five dementia patients and a quarter manager.

2 State of the Art

The state of the art of technological components for the assistive system enabling the quarter manager is discussed in this section.

2.1 Detection of Wandering

In order to discuss systems that detect wandering, it is useful to distinguish between two different approaches. In the first approach the person affected has to carry a device with an integrated GPS sensor. The GPS sensor may be integrated into a watch or a mobile phone. Several studies show that these devices may help to find or navigate wandering people [8, 9]. However, a disadvantage of GPS-based systems is that they have to be carried whenever the user leaves home. Dementia patients are likely to forget these systems at home.

The second approach is to place sensors in the flat of the patient and to inform the caregivers if wandering is detected. This approach can improve the sleep quality of caregivers [10] and help to prevent the diseased person to leave the house at inappropriate times if the caregiver is able to stop the person before he or she leaves. However, the sensor-based approach does not help outdoors. Furthermore it is necessary that a caregiver lives in the same accommodation or in the close proximity.

As shown, both approaches have disadvantages. It seems that currently available systems are only capable of wandering detection in specific surroundings.

2.2 Detection of Abnormal Behavior

A lot of research has been conducted to evaluate sensor setups and algorithms to detect abnormal behavior. The studies presented below have been selected because their approaches do not need the dementia patient to interact with the system. In addition, setups using cameras or microphones to monitor residents are considered dubious both from an ethical and a user acceptance perspective by the authors and were, therefore, not included in this section.

Steen et al. [11] described a system to model individual behavior using motion sensors. The user behavior was modeled based on timeslot and duration. In the timeslot-based model the probability of being present at a certain location at a certain time was computed. The duration-based model computed the probability of being present at a specific location with certain duration.

Lotfi et al. [12] used motion sensors and door sensors in their study. The sensor data was used to train different recurrent neural network approaches. As a result, the Echo State Network was identified as the most promising technique.

Wilken [13] identified activities by the data collected from a power sensor connected to the main fuse of an apartment. He defined activities as sequences of activations/deactivations of appliances. In his work, Wilken did recognize patterns of repetitive activities, but abnormal behavior patterns were not recognized by his system.

Taipa et al. [14] tagged domestic appliances using RFID technology to gain information about the activities of daily living. Their study showed that they were able to detect activities like toileting, bathing and grooming.

Most systems presented in this chapter are based on binary sensor data using simple home automation sensors. The advantage of these sensors is that they are cheap and easy to install. The presented systems are able to detect patterns of activities and abnormal behavior. However, they do not evaluate whether the detected anomalous behavior is relevant. Furthermore, these systems do not include information from caregivers (e.g. about expected behavior considered normal for a certain patient).

2.3 Stove Deactivation Systems

In case a person with dementia forgets to turn off the stove, fire may break out. To prevent this, stove detection systems are used. There are several commercial systems available which are described below. Many systems use a heat detector to monitor the temperature above the stove [15–17]. These systems deactivate the stove in case a predefined temperature is exceeded.

Another stove deactivation system includes an activation button [18]. If the user presses the button the stove is activated for an adjustable amount of time. After that time, the stove is turned off automatically. If the person wants to continue to use the stove, he or she has to press the button again. This system has a "comfort version", that uses a motion detector instead of the button.

2.4 Emergency Alert System

Emergency alert systems are used to ensure a stable connection between a household and an emergency alert center. In this subsection the emergency alert system of the Johanniter-Unfall-Hilfe [19] is described. This emergency alert device is connected to the telephone station. It contains an alarm button. If this button is pressed the device establishes a telephone connection to the emergency alert center and the resident can speak with an employee over loudspeakers. This employee assesses the situation and organizes help if necessary.

The emergency alert system has several additional features: Firstly, the alarm button can be used wirelessly. This enables the resident to set up an emergency call from every room in his apartment. Secondly, smoke and flood detectors can be connected to the emergency alert device. If these detectors transmit an event to the alert device, an automatic alarm will be transmitted to the emergency alarm center. Thirdly, a second button is placed on the emergency alert device. This button resets an internal alarm clock. If the button is not pressed for a predefined amount of time, an alarm call will be triggered.

3 Definition of Requirements

In this section the requirements definition for a quarter-based care concept is presented. To determine the needs of dementia patients and their relatives a workshop with caregivers (N = 15), interviews with relatives (N = 12) and interviews with experts (N = 5) have been conducted. Furthermore, the requirements for the ambient assistance system have been determined using the information provided by the caregivers, relatives and experts. The results of the workshop and the interviews are presented in the following section. In the second part of this chapter the requirements for the quarter-based care concept are described, based on the caregiving principles in assisted living communities.

3.1 Requirements Derived from the Workshop and the Interviews with Relatives and Experts

As a first step of the requirements analysis, a workshop with professional caregivers was conducted. In the workshop, the "world café" method which is a good method for structuring large group discussions [20] was used. The four topics were:

- the own house—technical assistance opportunities,
- interconnections in the quarter,
- interconnections of caregivers and their services,
- individual situation—problems and obstacles with supplying dementia patients, wishes and ideas for the future.

As a general result it can be said that dementia patients vary significantly in their needs. Therefore, any technical assistance system in the flat has to be modular und adjustable to individual (and perhaps changing) needs of the patient. Another finding was that people who suffer from dementia are not able to learn any new interactions with technical systems. Thus, the patients should not have to interact actively with technical systems in any manner and known appliances have to remain in use as long as possible. Furthermore, a need for better connecting relatives, physicians and caregivers was reported. Finally, it could be observed that a sensitization to the needs of all actors involved in the quarter is necessary. These actors include not only the caretakers and the quarter manager, but also physicians, pharmacists, local stores, banks, police stations, etc.

After the workshop, relatives of dementia patients were interviewed. For this purpose a questionnaire was designed to serve as a guideline. Each interview lasted about 1 h. In the interviews the relatives talked about their experience with their caretaking of dementia patients, especially in the early phase of the disease. Because most of the patients in question had already been suffering dementia for a long time (4–17 years, average 8.6 years), the relatives had to recall their experience from that time. The patients in question were in the age from 63 to 90 years.

Table 1 Symptoms reported in the survey and their frequency (N = 12, multiple answers allowed)

Symptom	Frequency
Emotional abnormalities (Lustlosigkeit, Distanzlosigkeit, Aggressivität, …)	6
Problems with the handling of money	4
Disorientation	3
Wandering	3
Nutrition problems	2
Misjudgment of daytime	2
Forgetting to turn off the stove	2
Failure to correctly assign household articles to their functions	2

The majority of them had additional diseases and needed several hours of care each day. Although the number of patients discussed during the interviews is small (N = 12), some general conclusions can be made.

Generally, it was confirmed that dementia patients have a diversity of needs, which was also a finding of the workshop with professional caregivers. To exemplify this, the following table lists a range of symptoms which were reported by relatives answering the question: "Which symptoms of dementia made it impossible or very difficult for your relative to stay in his/her own apartment?"; it was possible to report more than one symptom.

The symptoms reported in Table 1 allow for several conclusions. Firstly, eight different symptoms were reported for only twelve people. Secondly, no symptom was reported in more than 50% of the interviews. Thus, the diversity of dementia patients is confirmed. The impression of this diversity becomes even sharper if the combinations of the mentioned symptoms are taken into account. Every answer was unique in terms of the combination of symptoms.

The diversity of the dementia patients was not the only finding gained from the interviews. It was confirmed that people with dementia are incapable of learning how to operate new technical devices. Even one-button systems turned out to be too complicated for almost all of the dementia patients in question. In addition, the interviews helped to prioritize the problems related to dementia. Generally, it can be said that physical well-being and safety turned out to be more important for dementia patients and their relatives than social needs, such as participation and entertainment. Therefore, Maslow's "Hierarchy of Needs" [21] has been confirmed for the patients in this study, at least from the perspective of the relatives interviewed.

The interviews also showed that the relatives have a great need for information and organization. The majority of them expressed that they had not known about local care possibilities like occupational therapy or group therapy for a long time.

Finally, some relatives mentioned that they were overstrained with planning the upcoming days for their diseased family members. There is a need for a supporting third party that has an overview about the schedule of the patient, the relatives and the caretakers. In the QuoVadis project, this third party will be the quarter manager.

As the third step in the requirements analysis, experts were interviewed. This group consisted of a gerontologist, two department leaders from the city government of Oldenburg and two leaders of care facilities. In general, they confirmed the findings of the workshop and the interviews. In addition, it was mentioned that people with dementia are often also affected by other diseases. This adds to the finding that dementia patients are very diverse and need individual support. Finally, housing counselling was mentioned as useful non-technical method to reduce the risks of falls and disorientation.

3.2 *Study of Assisted Living Communities for Dementia Patients*

To determine effective care principles and assistance opportunities for technical systems, existing assisted living communities for demented inhabitants were evaluated. For this purpose, an assisted living community in Oldenburg, Germany was visited and the person in authority for the community was interviewed. Furthermore, two people with relatives in assisted communities were consulted. It was examined how the guidelines for caregiving for dementia patients in Germany [22] are applied in assisted living communities. Several interesting insights could be gained.

Firstly, there is always a carer or caretaker in the assisted community who can conduct measures if needed. Thanks to this, problems can be addressed immediately at any time of the day. In the QuoVadis project this 24/7 presence of a caretaker cannot be translated directly into the quarter based setup. This has to be addressed by a technology-based security concept that is presented in Sect. 4.1.

Secondly, the caregivers and carers in assisted living communities are well aware of the individual needs of every resident. Therefore, they are able to organize necessary measures for the patients. Such measures can be occupational therapy, physical activities, art therapy, etc. The aim of the QuoVadis project is to realize individual care organization as described above. Therefore, technical assistance will be necessary since the quarter manager is not with the dementia patient all night and day and, therefore, not fully aware of all individual needs. The technical assistance concept to meet this requirement is presented in Sect. 4.2.

Thirdly, the inhabitants of the assisted living community participate in activities of daily living of the inhabitants, such as cooking. This participation structures the day and makes the inhabitants feel accepted and useful. In a similar manner, the quarter manager will organize group activities for the inhabitants in the quarter.

Fourthly, the relatives are relieved with respect to the organization effort, because they do not have to organize the daily care for their demented relative on their own. This does not mean that the relatives are not involved in the care process in assisted living communities. On the contrary, the caretakers strongly rely on the help from the relatives. In the QuoVadis project it can be the quarter manager who relieves the relatives in organization. Similar to the assisted living community the quarter manager will need the help of the relatives to ensure good care for the residents.

4 QuoVadis System Architecture

This chapter describes the derived QuoVadis system architecture. The focus is mainly on the technical aspects of the concept. The technical concept can be divided in two systems, or modules: (1) the security system and (2) the individual care system. The security system addresses the needs of safety and security that have been found to be most important to dementia patients and their relatives. The individual care system will enable the quarter manager to organize individual care measures for the affected people. Furthermore, the individual care system will inform the quarter manager about abnormalities, based on the feedback to previously reported abnormalities and additional information that the quarter manager has reported to the system.

4.1 Security

The security system consists of three subsystems: an emergency alert system, a wandering detection system and a stove-deactivation system.

The emergency alert system has been designed such that residents who are still able to learn how to operate an alarm button are enabled to call for help this way, whereas autonomous ambient sensors are integrated into the system for users unable to remember the alarm button.

The structure of the emergency alert system is shown in Fig. 1. Every flat of the participants in the project will be equipped with an emergency alert device. On the input side this device receives information from sensors in the flat of the dementia

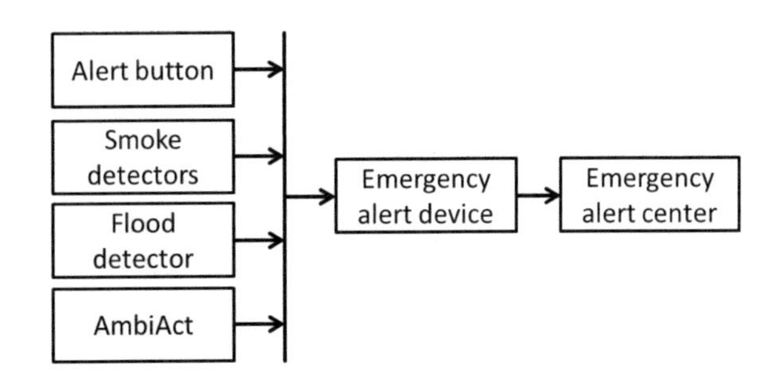

Fig. 1 Emergency alert system

patient. The wirelessly connected sensors are the alert button, which can be pressed by the inhabitant in case of emergency, smoke and flood detectors that autonomously raise alarms in case of emergency and the AmbiAct. The AmbiAct is a smart power sensor that detects the usage of electrical appliances [23]. If the usage of an appliance is detected, the AmbiAct sends a signal to the emergency alert device, which in turn resets an internal alarm clock. If the alarm clock is not reset for a configurable period of time (normally 24 h), the emergency alert device generates an alarm signal because of unexpected inactivity in the flat.

On the output side, the emergency alert device connects with the emergency alert center. The emergency alert center is called in case one of the sensors transmits an alarm or the AmbiActs do not reset the internal clock before the configured timeout. In case the emergency alert center receives an alarm call, an employee tries to contact the inhabitant by telephone. If it is not possible to contact the resident, the employee calls for help to come to the apartment of the dementia patient.

The wandering detection subsystem is designed to detect wandering of the resident outside of his or her usual daytime. At first, a system to learn the circadian rhythm of the resident will be realized. For this purpose, motion sensors and a power sensor in the main fuse of the flat will be installed. Using the data of these sensors, the circadian rhythm can be estimated. Furthermore, to detect wandering a wireless contact sensor will be placed at the entrance door. The contact sensor transmits events to a computer that serves as core element for the assisted living system in the flat of the dementia patient. If the contact sensor transmits events at times beyond the learned circadian rhythm, a message will be sent from the computer to the emergency alert center. This alarm will be handled by the employees in the alert center in the same manner as the alerts from the emergency alert device.

However, the system presented does not include any assistance to localize a wandering person once he or she has left the home. Since there are currently no systems available that would not require the user to carry a GPS transmitter, no wandering detection or localization will be used outdoors. To prevent people from getting lost, a sensitization of the actors in the quarter is planned that will increase the probability that a wandering person receives help.

The stove deactivation system will prevent the stove from staying activated if the resident forgets to turn it off. Although some commercial products exist a different solution needs to be developed for the QuoVadis project. The disadvantage of commercial stove deactivation system is that they are stand-alone systems that cannot be connected to our assistive system. In order to inform the quarter manager whether the stove is not turned off regularly, a connection between stove deactivation and the computer is necessary. Hence, a stove deactivation system will be developed in the project.

Finally, counseling on housing will be provided to the residents to lower the risks of falls and disorientation in their own apartment.

4.2 Individual Care

The individual care system addresses long-term changes of the behavior and the needs of the dementia patient. It can be divided into a technical component, called the Adaptive Care Support System (ACSS), and an organization and activation effort.

4.2.1 Adaptive Care Support System

The adaptive care support system (ACSS) will enable the quarter manager to organize and optimize personalized support for the dementia patients in his or her quarter. This system does not require the residents themselves to interact with technology in any manner. The aim of the ACSS is to use information from the quarter manager along with information on activities that have been recognized by technical systems to identify and predict behavior problems. The ACSS contains four subsystems, which are presented in Fig. 2.

The first subsystem is denoted as the sensory subsystem. This system consists of multiple wireless sensors that are connected to a computer in the flat of the resident. A power sensor will be placed in the main fuse to detect the use of electrical appliances. In order to identify the activation and deactivation of the appliances, supervised machine learning will be used. Once the identification of appliances is known only one sensor is necessary to monitor the usage of all electrical devices. This is a great advantage of this sensor type. Sensors to monitor the water consumption will be used in the bathroom and the kitchen of the apartments. Additionally, motion sensors will be placed in every room to monitor the movement of the inhabitant. The sensors in this setup communicate wirelessly with the computer in the apartment. All the used sensors are unobtrusive as this was a requirement defined in the previous section. The events of all sensors in the apartment will be mapped into the sensor data vector ($\vec{S}$). This vector will be used in the classification system to determine activities of daily living. The following activities of daily living should be recognized by the classification system:

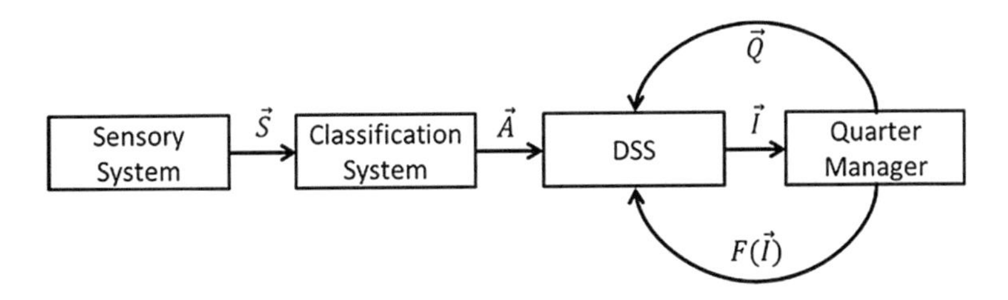

Fig. 2 Adaptive care support system

- eating,
- cooking,
- toileting,
- grooming.

Furthermore, the activity (movement) of the resident in the flat will be estimated. Since the QuoVadis project focuses on the interaction between quarter manager and the technical system it is not the aim to improve algorithms for the detection of daily activities. Therefore, an existing algorithm will be used in the classification system to determine the activities of daily living. To identify a suitable algorithm, a pilot study will be conducted. This algorithm will have to be more complex than most of the solutions presented in this chapter because the data at the input of the classification vector is generated by different non-binary sensors. Once an algorithm to identify activities is selected and implemented, the classification system will report the recognized activities in an activity vector ($\vec{A}$). Additionally, the duration and the daytime of each activity will be stored in this vector.

This information from the sensory system will then be transmitted to the third subsystem of the ACSS, the Decision Support System (DSS). The DSS is the link between the quarter manager and the technical system. It accumulates activity information ($\vec{A}$) from the technical systems and from the quarter manager ($\vec{\mathbf{Q}}$). The quarter manager reports information on activities that have taken place outside of the apartment (i.e. walks, meals, visits, etc.) and his or her observations regarding the overall health of the patient. This information is mapped in the second input vector ($\vec{\mathbf{Q}}$). Based on these and additional information from the quarter manager, the DSS is autonomously able to inform the quarter manager when relevant changes in behavior occur.

The quarter manager receives an information vector ($\vec{I}$) if the DSS detects abnormal behavior. To enhance the quarter manager to understand the information vector ($\vec{I}$), this vector needs to be presented in an understandable way. Subsequently, the quarter manager verifies the received information by contacting the dementia patient and reports his feedback on this information ($F(\vec{I})$) to the DSS. This feedback can be one of the following statements:

- positive: the abnormality reported by the DSS is relevant for the quarter manager;
- positive, but based on exception: the abnormality reported is relevant, but is based on exceptional circumstances (i.e.: patient is on vacation, patient is visited by many people and therefore a lot of movement is registered);
- negative: the abnormality reported by the DSS is not relevant for the quarter manager;
- false: the abnormality reported by the DSS is based on technical errors. Therefore, the related data should not be used for further predictions.

The feedback from the quarter manager is used by the DSS for future predictions.

4.2.2 Organization and Activation

In knowledge of the information gained by the ACSS, the quarter manager can provide individual activation measures for the patient. These activation measures can be walks with volunteers, shared meals, group activities and many more.

To relieve the relatives in the organization of daily activities, an individual calendar will be implemented for each resident who is involved in the quarter care concept. The calendar is accessible by the residents, their relatives and the quarter manager. Additionally, some information of the resident will be stored in the calendar (name, age, photo, medication etc.). This information can be accessed by the quarter manager in case of emergency.

5 Conclusion

This article presented the requirements and the system architecture for the Quo-Vadis project. So far, this general concept has been discussed with two quarter managers, who confirmed its utility. In addition, the concept will be evaluated in a workshop with professional caregivers and interviews with experts.

In the first half of 2016 the technical systems will be implemented. The DSS and the infrastructure for the communication between a professional caregiver and a technical system will be developed. It will be examined how the information vector can be transformed into understandable information for the quarter manager. Similar to this, a method to map the feedback and the information from the quarter manager in the feedback and information vectors will be determined. Furthermore, sensitization of the quarter will be conducted by this time.

A pilot study to evaluate the technical systems will be conducted in the second half of 2016. The data gained in this study will help to compare different machine learning algorithms for the DSS. In addition, the detection of activities will be tested at that time.

The start for the realization of the project is planned for the beginning of 2017.

Acknowledgements This work was funded by the Central Federal Association of the Health Insurance Funds of Germany (GKV-Spitzenverband).

References

1. Kremer-Preiß, U.: Wohnatlas—Rahmenbedingungen der Bundesländer beim Wohnen im Alter. Kuratorium Deutsche Altershilfe/Wüstenrot Stiftung, Teil 2: Zukunftsträchtige Strategien im Politikfeld Wohnen im Alter, p. 8 (2014)
2. Ferraro, K.F.: Relocation desires and outcomes among the elderly. Res. Aging **3**(2), 166–181 (1981)
3. Payk, T.R.: Demenz, Ernst Reinhardt Verlag München Basel 2010, p. 45 (2010)

4. Beerens, H.C., et al.: Change in quality of life of people with dementia recently admitted to long-term care facilities. J. Adv. Nurs. **71**(6), 1435–1447 (2014)
5. Dankers. L et al.: Leben Demenzkranke zuhause länger als im Heim? Zeitschrift für Gerontologie und Geriatrie 4 pp. 254–258. Springer, (2010)
6. Sonntag, K., von Reibniz, C.: Versorgungskonzepte für Menschen mit Demenz, p. 11. Springer, Berlin Heidelberg (2014)
7. Kremer-Preiß, U.: Wohnatlas - Rahmenbedingungen der Bundesländer beim Wohnen im Alter. Kuratorium Deutsche Altershilfe/Wüstenrot Stiftung, Teil 1: Bestandsanalyse und Praxisbeispiele, p. 32 (2014)
8. Lin, Q. et al.: Detecting wandering behaviours based ond gps traces for elders with dementia. In: International Conference on Control, Automation, Robotics and Vision (2012)
9. Sposaro, F. et al.: iWander: an android application for dementia patients. Engineering in Medicine and Biology Society (EMBC).In: 2010 Annual International Conference of the IEEE. IEEE (2010)
10. Spring, J., et al.: Improving caregivers' well-being by using technology to assist in managing nighttime activity in persons with dementia. Res. Gerontol. Nurs. **2**(1), 39 (2009)
11. Steen, E., et al.: Modeling individual healthy behavior using home automation sensor data: results from a field trial. J. Ambient Intell. Smart Environ. **5**, 503–523 (2013)
12. Lotfi, A., et al.: Smart homes for the elderly dementia sufferers: identification and prediction of abnormal behavior. J. Ambient Intell. Humaniz. Comput. **3**(3), 205–218 (2012)
13. Wilken, O.: Aktivitätserkennung basierend auf Nutzung elektrischer Geräte im häuslichen Bereich. Verlag Dr, Hut (2013)
14. Tapia, et al.: Activity recognition in the Home Using Simple and Ubiquitous Sensors. Springer, Berlin Heidelberg (2004)
15. locate solution!: my.cook.guard. http://www.locatesolution.de/produktloesungen/my.cook.guard.html. Accessed 10 Nov 2015
16. Deutsches Rotes Kreuz: Automatische Herdabschaltung. http://www.drk-karlsruhe.de/angebote/senioren/herdabschaltung.html. Accessed 13 Nov 2015
17. Indexa: Herdwächter. http://www.indexa.de/w2/re_brandschutz_herdwaechter_herdalarm.htm. Accessed 13 Nov 2015
18. PIC-TEC: Herdüberwachung mit Tasterbedienung. http://www.pic-tec-shop.de/Firmware-p85h77s78-Herdueberwachung-HA1.html. Accessed 13 Nov 2015
19. Johanniter-Unfall-Hilfe e.V.: Hausnotruf Auf Draht! Kunden-Information zum Johanniter Hausnotruf. http://www.johanniter.de/fileadmin/user_upload/Dokumente/JUH/BW/Factsheets/HNR/RV_S/Infomaterial_HNR_Mai_2010.pdf. Accessed 13 Nov 2015
20. Brown, J. et al.: The world café, systems thinking and chaos theory. Netw. Newslett. Jan. (2002)
21. Maslow, A.H.: A theory of human motivation. Psychol. Rev. **50**, 370–396 (1943). http://www.researchhistory.org/2012/06/16/maslows-hierarchy-of-needs/. Accessed 09 Nov 2015
22. S3-Leitlinie Demenzen (Kurzversion), Deutsche Gesellschaft für Psychiatrie, Psychotherapie und Nervenheilkunde (DGPPN), Deutsche Gesellschaft für Neurologie (DGN), Nov 2009
23. Iatridis, K.; Schroeder, D.: Responsible Research and Innovation in Industry, pp. 15–16. Springer (2016)

Part II
Technology for Smart Environments

Invisible Human Sensing in Smart Living Environments Using Capacitive Sensors

Andreas Braun, Silvia Rus and Martin Majewski

Abstract Smart Living environments aim at supporting their inhabitants in daily tasks by detecting their needs and dynamically reacting accordingly. This generally requires several sensor devices, whose acquired data is combined to assess the current situation. Capturing the full range of situations necessitates many sensors. Often cameras and motion detectors are used, which are rather large and difficult to hide in the environment. Capacitive sensors measure changes in the electric field and can be operated through any non-conductive material. They gained popularity in research in the last few years, with some systems becoming available on the market. In this work we will introduce how those sensors can be used to sense humans in smart living environments, providing applications in situation recognition and human-computer interaction. We will discuss opportunities and challenges of capacitive sensing and give an outlook on future scenarios.

1 Introduction

The selection of sensors for smart environments has been constantly increasing in the last decades [1]. There is a large variety of off-the-shelf sensors that detect movement, temperature, presence, air quality or other activities [2]. Particularly in the last two years some high-profile acquisitions have shown the interest of large IT companies such as Google and Samsung to have a more active role in the field of smart homes [3, 4]. The notion of Ubiquitous Computing or Ambient Intelligence has envisioned ensembles of sensors that are connected and seamlessly integrate into our everyday environment [5]. However, while progress has been made in numbers and connectivity, the sensors still very much look like appendages in our living environments. Streitz et al. proclaimed the notion of the disappearing computer that moves away from our direct view and is only exposed in terms of its input

A. Braun (✉) · S. Rus · M. Majewski
Fraunhofer Institute for Computer Graphics Research IGD, Darmstadt, Germany
e-mail: andreas.braun@igd.fraunhofer.de

R. Wichert and B. Mand (eds.), *Ambient Assisted Living*,
Advanced Technologies and Societal Change,
DOI 10.1007/978-3-319-52322-4_3

and output channels [6]. This has become very true for devices, such as smartphones and tablets, but remains a challenge for sensor systems.

There has been considerable research in this domain. A first example is the radio-frequency heart rate and breathing monitor by Adib et al. [7]. Remote antennas can be used to analyze the radio waves reflected from a human body and apply signal processing that detects very fine variations in movement. This is sufficient to gather physiological parameters of a human body at a distance of several meters without requiring any wearable system. Another system based on wireless signals is WiSee by Pu et al. that uses signal reflections in the 2.4 GHz band to detect gestures performed by a standing user [8]. Radiofrequency sensing is an example technology that does not require any contact to the user and additionally works through many different materials. The sensors could be integrated into a wall or other features of the environment. They are one example for an invisible sensor.

Invisible sensors have two main characteristics: (1) their sensing methods are not perceivable by the human senses of perception, we can't feel them being used and (2) they are not exposed to the environment, thus being invisible to the human eye. This seamless integration of these invisible sensors has the potential to revolutionize smart living environments in terms of feasibility, functionality, and acceptance. Apart from the mentioned examples of radiofrequency sensors there are other systems that support this invisible sensing. In this work we want to focus on capacitive sensors that use weak electric fields and can detect the presence of human bodies over the distance. We will show, how they can be regarded as invisible sensors in Smart Living environments and give an overview of existing applications in previous works of research.

On the following pages we will give a brief introduction to capacitive sensing and its common applications. We will outline two application domains for Smart Living—activity recognition and interaction in smart environments, presenting example systems and common data processing methods. We will identify and describe several use cases in Activity Recognition, Interaction Devices and Smart Objects, and finally discuss opportunities and challenges, as well as giving an outlook on future developments.

1.1 Capacitive Sensing

The human body is mostly composed of ionized water and differences in the proportion of water in specific types of body tissue are causing distinct electrical properties that can be distinguished. Thus, we can either use external electric fields and measure the influence of the human body moving within, or couple the body to an electric transmitter and measure the resulting field. An early application is the electronic musical instrument Theremin, developed by Leon Theremin in the 1910s [10]. A classic work in the field of capacitive proximity sensing is the dissertation "Electric Field Imaging" by Joshua Smith that aggregates the works performed

Fig. 1 Electrode materials samples of same size used in self capacitance measurement mode: (from *left* to *right*) copper electrode, conductive paint, conductive thread, conductive fabric, conductive paint on fabric [9]

using capacitive proximity sensors by the MIT Media Lab in the 1990s, including their introduction of different measurement modes for capacitive sensing [11].

The available measurement modes are self-capacitance and mutual capacitance measurement mode. In self capacitance measurement mode, one receiver electrode is needed, whereas in mutual capacitance measurement mode two electrodes, a transmitter and a receiver electrode are needed. In self capacitance mode the change of electric field between the electrode and the conductive body or object is measured, whereas in a mutual capacitance setup the change of the electric field as a result of a body or conductive object being placed between the two electrodes is measured. In most of our own developed applications [12–14] and the further related works [15, 16], where capacitive sensing is not used to determine touch, the self-capacitance measurement mode is more prominent due to the fact that it needs only one electrode. Looking into the area of flexible electrodes, the self-capacitance measurement mode is also more widespread [17–19].

In several previous works we evaluated different materials and shapes for capacitive sensing in smart environments [9, 20]. Regarding electrode size it should be related to the desired application. If the application is not restricting the available space, the electrode should be approximately of the same size as the object that has to be detected. This generates the highest difference in capacitance when the distance is changing. With regards to material, we found that most conductive materials behave similar in all measurement modes. However, there were some outliers in transparent conductive polymers that showed unfavorable properties [21]. For various different flexible electrode materials, it was found that without the influence of noise all electrode materials would be equally well suitable for flexible capacitive sensor applications [9] (Fig. 1).

Given the existing literature, we identified two general aspects of Smart Living that can be covered by invisible human sensing systems. The first is activity recognition that benefits from additional sensing systems in the environment, the second is interaction in smart environments, whereas the invisible sensors are used to create interactive surfaces or zones, used to control aspects of the Smart Living environment. In the following two sections we will give a very brief introduction to the two before detailing our work that has been done in the previous years.

1.2 Activity Recognition

In the scope of Smart Living environments, activity recognition is one of the most important technological components. The idea is to detect typical activities of daily living (ADLs), in order to provide suitable and targeted assistance. These are structured events that can be distinguished and classified. A seminal work was done at the MIT in the early 2000s that recorded and classified various ADLs using simple sensor systems [22]. However, there are also different approaches to detect these kinds of activities, e.g. using wearable sensors. Lara and Labrador provide a good overview of technologies and methods used in this domain [23]. They evaluate 28 systems and provide a taxonomy of learning methods and classification.

Most systems rely on machine learning approaches to detect activities. Several features in either the time- or frequency-domain are extracted from the raw sensor data and further analyzed using some form of learning. The features may be coming from a single sensor or multiple systems using methods of sensor data fusion [24]. We can further distinguish supervised and non-supervised learning methods. In the first, the features are associated to an activity by humans or specialized systems, in the second, the learning method tries to find significant features that can be associated to activities that are grouped together and can be tagged in a later stage of the process.

Invisible sensors can be integrated into everyday objects, in order to recognize ADLs. Our work focuses on invisibly integrating capacitive sensors into furniture in a Smart Environment, e.g. in the chair, in the bed or the table. Using the furniture's existing materials as electrodes, or integrating i.e. interwaving the electrodes into them, the sensors become completely concealed by the furniture's surfaces. The sensors become an unobstrusive part of the environment. However, being installed as part of the environment, the sensors have the drawback of not being directly attached to the object of interest. Nevertheless, having a large number of invisibly embedded sensors surrounding the object of interest can still deliver the required information, as adequate as having a small number of sensors attached directly and it enables the system to recognize ADLs of several objects without the need of additional object-attached sensors. Capacitive sensors are especially well suited for being deployed in a large number, since they only consume a small amount of energy and require relatively low original costs.

In order to overcome some recognition limitations and to improve the unobtrusiveness of systems using other sensing methodologies, sensor fusion can be used.

1.3 Interacting in Smart Environments

The second area that can be covered by invisible sensors are interaction systems for Smart Living environments. These encompass all methods that are used by humans for control. These can be regular interaction devices, such as mouse and keyboard, or touch devices including smartphones. There are additionally specialized interaction systems, e.g. for gestural control, such as the Microsoft Kinect, or the Leap Motion [25, 26]. However, interaction systems can also be integrated in a way that they are integrated into everyday objects. This concept is often called interactive surface. Examples include sensor-augmented tables, or sensor devices that can detect interaction on an attached object [27, 28]. A challenge in this regard is the feedback to the user if any interaction successfully occurred, leading to concepts, such as projectors that can be freely moved, or laser systems that indicate pointing locations [29–31].

1.4 Applications in Activity Recognition

The Smart Bed is a cloth sheet with a mutual capacitive sensing wire grid that can be placed above the mattress of a bed in order to detect postures of a person [32]. Detecting the type and duration of postures is particularly important for intensive care, as prolonged lying can cause decubitus wounds. The system detects five different lying positions with an accuracy of more than 90% using a nearest-neighbor classification. Figure 2 shows on the left the resulting normalized sensor values of a heavy person sitting on the smart bed. The system could alert the carers, when a person is lying in a certain position for too long. This can be done by visualizing the course of detected lying position changes. With this information, carers can decide upon the next action to perform.

The second system is the Capacitive Chair used for posture recognition and working activity detection, e.g. to be applied in a home office, as shown in Fig. 3 on the left [13]. It uses eight self-capacitance sensors that are placed under the seat pan, on the arm rests and in the back rest. The postures can be used to encourage persons to regularly change their posture or support exercises on the chair, as shown in a later system [12]. The accuracy of detection with four postures was higher than 98%. The work activity recognition used the intensity of movements over a period of time to distinguish between persons not on the chair, working inactively (e.g. reading from screen), or actively (e.g. engaged phone conversation). In addition, the system can be used to recognize the breathing rate. The large sensor in the center back will detect the movement of the chest during breathing. A similar setup was placed inside a car seat demonstrator, as shown in Fig. 3 on the right [33]. The posture recognition is used to adjust the seat to the driver's specific physiological needs supporting long term usage during a journey. Safety aspects are supported by

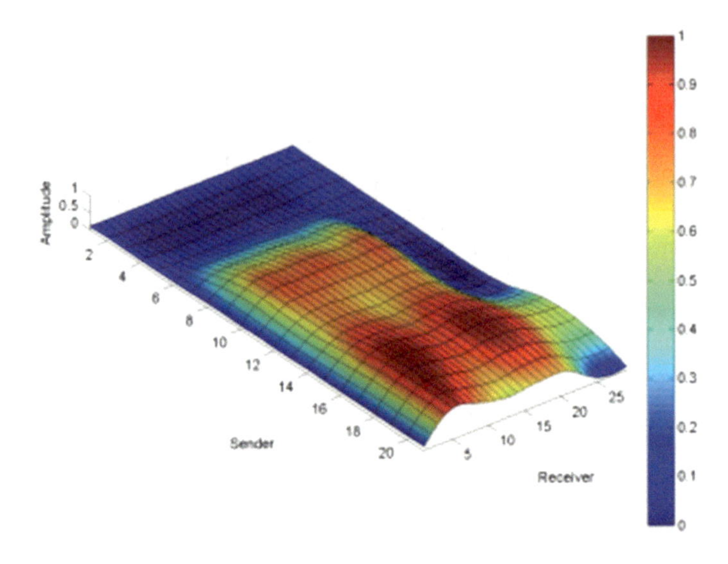

Fig. 2 Sensor response of person sitting on the smart bed [32]

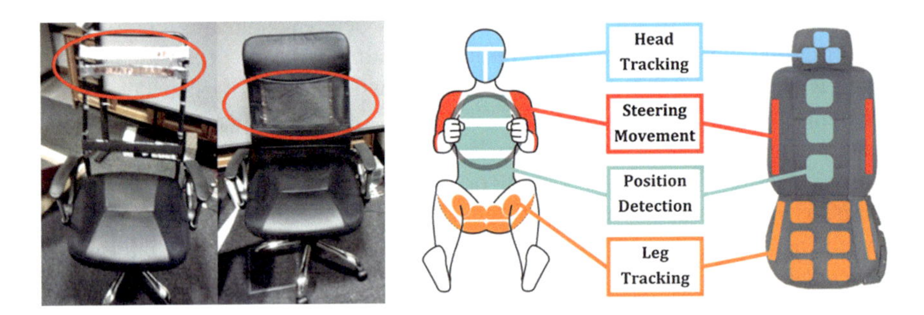

Fig. 3 *Left* Capacitive Chair prototype. *Right* Automotive seat with four areas of interest

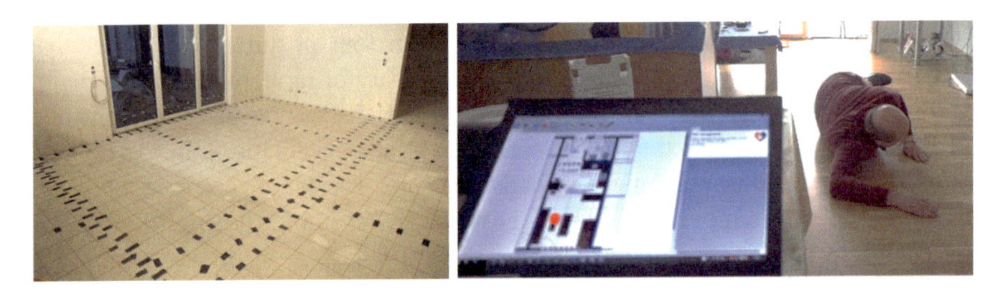

Fig. 4 *Left* CapFloor setup inside a new construction. *Right* Example of the fall detection functionality

the activity recognition. The system recognizes e.g. potential microsleep situations, by analyzing the driver's head movements, i.e. recurring nod of the head.

The third system is CapFloor, an indoor localization and fall detection system that uses a grid of electrodes under a floor layer and sensors that are placed at the walls [34] (Fig. 4). The system can distinguish multiple users and detects lying persons by analyzing the size of the detected object. A potential use case is application in retirement homes, in order to control energy saving and comfort measures, or to alert care personnel of a fall that happened.

1.5 Applications in Interaction with Smart Environments

Two example interaction devices will be presented in this section. The first is CapTap, a capacitive interaction table that uses 24 self-capacitance sensors to track the movement of two hands above the surface—allowing mid-air gestural interaction [35]. Additionally, there is a microphone attached that performs acoustic touch detection and distinguishes several different types of events. A view of the inner part of CapTap is shown in Fig. 5 on the upper right. We can see the copper electrodes and attached to those the small sensor boards. The microphone is placed in the center of the surface (Fig. 5).

Curved, the second interaction device presented, is an interaction system that is intended to ergonomically follow the natural movement of hands [14] (Fig. 6). It is based on capacitive proximity sensors that are placed below a free-form surface that track hand movement at a certain distance. In a first iteration this was used to

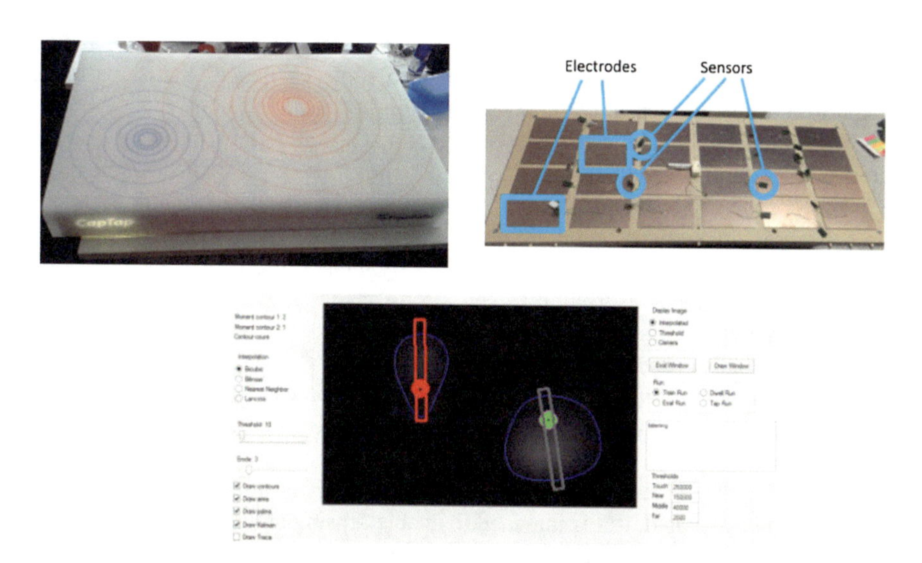

Fig. 5 *Upper left* CapTap Device—*Upper right* Electrodes and sensors setup—*Bottom* Multiple-hand gesture recognition visualization

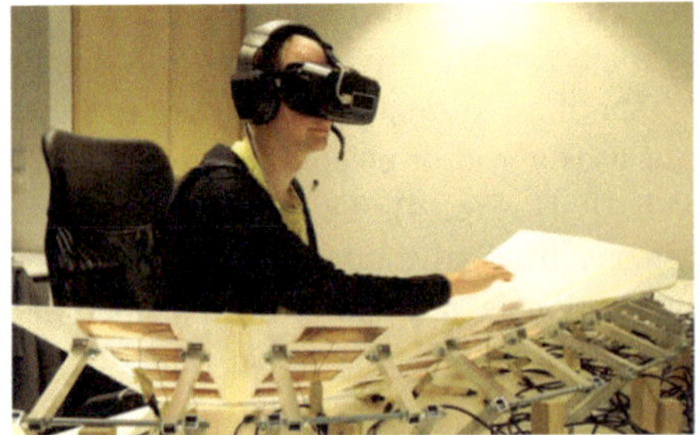

Fig. 6 *Left* Curved device implementing free-form capacitive sensor surfaces—*Right* Natural curvature supports interaction in Virtual Reality scenarios

simplify interaction in virtual reality applications. However, there are potentially more interesting applications in Ambient Assisted Living. For examples the concept can be extended to create tailored interaction systems for persons with motor deficiencies that are suited for their individual motor range. Furthermore, a dynamic adaptation to changing postures and movement ranges is in the developing state. The Curved-surface changes its shape and spatial position to guarantee the optimal distance to the user's extremities, optimizing the signal to noise ratio, the working range of the sensors and the usability.

1.6 Opportunities and Challenges

The notion of invisible sensing in Smart Living environments using capacitive sensors has a number of opportunities but also some challenges that have to be taken into account when building an application in such a system.

Any application can be realized invisibly using capacitive sensors, as long as the electrode is only covered by non-conductive materials. Unobtrusive application, e.g. using conductive threads is equally promising. The sensing technology is not prone to certain environmental influences, e.g. light. It is therefore feasible to replace cameras in settings where strongly varying lighting conditions are to be expected. Capacitive sensors don't need strong processing. As each sensor only delivers low-byte values several hundred times a second, small microprocessors are able to handle most processing tasks. Additionally, this also serves as a contributing factor to preserving privacy, as only limited information can be acquired about the user. Finally, the energy consumption is very low, allowing battery use over prolonged periods of time.

As for the challenges—the resolution of the sensors is limited. Only a limited number can be installed, if it is intended to enable sensing at larger distances. Small electrodes only allow for a limited range. Therefore, less data is acquired and smart methods have to be used in order to enable activity recognition and interaction.

While light and sound do not disturb the measurement, some other environmental factors may do so. Temperature and humidity change the properties of an electric field and have to be compensated for. A larger challenge are radio-frequency systems in the environment that may emit waves in ranges that make the electrodes act as antennas. This has to be taken into account during sensor design, using appropriate filtering methods. Finally, proximity sensors are not particularly good at detecting touches, unless the electrode is exposed. An open hand that is close to the electrode may have the same response as a finger that is touching the electrode. This has to be taken into account, when designing the electrode layout for any given application.

1.7 Outlook and Conclusion

On the previous pages we have introduced the notion of invisible capacitive sensors for human sensing in Smart Living environments. Appropriately applied, they are very well suited to create systems for activity recognition. We presented two systems that enable physiological sensing, including breathing rate and posture recognition in varying levels of detail. Two interaction devices have been introduced that can be integrated into common surfaces, or used to control novel forms of virtual reality interaction. Finally, capacitive sensors were also used to create Smart Objects—everyday items that are augmented with sensing technologies.

As future opportunities we would like to explore the options that arise when combining invisible sensing with ensembles of wearable devices. The combination of invisible sensing and data acquired near the body seems promising. Additionally, electric sensing can be used to augment sensor and processing nodes for the Internet of Things, e.g. by making them aware of nearby activities or the presence of a human body, which may trigger a range of events. Data fusion within smart living networks is another interesting option. While single invisible sensors may not reach the precision of e.g. cameras, smart fusion algorithms can help in alleviating this problem. Finally, we are investigating the integration of another category of sensors —the electroscope or electric potential sensors that detects changes in the ambient electric field created by charged objects, such as the human body, acting within.

References

1. Cook, D.J.: How smart is your home? Science (New York, NY) **335**, 1579 (2012)
2. Cook, D., Das, S.: Smart Environments: Technology, Protocols and Applications. Wiley (2004)
3. Google Inc.: Google to Acquire Nest—Investor Relations—Google. http://investor.google.com/releases/2014/0113.html. Accessed 28 April 2016)

4. Chen, B.X.: Samsung Acquires SmartThings, in Embrace of the Smart Home. http://bits.blogs.nytimes.com/2014/08/14/samsung-acquires-smartthings-in-embrace-of-the-smart-home/?_r=0. Accessed 28 April 2016
5. Weiser, M.: The Computer for the 21st Century. Sci. Am. **265**, 94–104 (1991)
6. Streitz, N., Nixon, P.: The disappearing computer. Commun. ACM **48**, 32–35 (2005)
7. Adib, F., Mao, H., Kabelac, Z., Katabi, D., Miller, R.C.: Smart Homes that Monitor Breathing and Heart Rate. In: Proceedings of the 33rd Annual ACM Conference on Human Factors in Computing Systems, pp. 837–846 (2015)
8. Pu, Q., Gupta, S., Gollakota, S., Patel, S.: Whole-home gesture recognition using wireless signals. In: Proceedings MobiCom. pp. 27–38 (2013)
9. Rus, S., Sahbaz, M., Braun, A., Kuijper, A.: Design factors for flexible capacitive sensors in ambient intelligence. In: Ambient Intelligence, pp. 77–92. Springer (2015)
10. Glinsky, A.: Theremin: Ether Music and Espionage. University of Illinois Press (2000)
11. Smith, J.R.: Electric field imaging. Ph.D. Dissertation, Massachusetts Institute of Technology (1999)
12. Braun, A., Schembri, I., Frank, S.: ExerSeat—sensor-supported exercise system for ergonomic microbreaks. In: Ambient Intelligence, pp. 1–16 (2015)
13. Braun, A., Frank, S., Wichert, R.: The capacitive chair. In: Proceedings DAPI (2015)
14. Braun, A., Majewski, M., Zander-Walz, S., Kuijper, A.: Curved—free-form interaction using capacitive proximity sensors (2015)
15. Haescher, M., Matthies, D.J.C., Bieber, G., Urban, B.: CapWalk : A Capacitive Recognition of Walking-Based Activities as a Wearable Assistive Technology, pp. 1–8
16. Kaila, L., Raula, H., Valtonen, M., Palovuori, K.: Living Wood—A Self-Hiding Calm User Interface, pp. 267–274
17. Janeiro, R. De Fuks, H.: Hairware: The Conscious Use of Unconscious Auto-contact Behaviors, pp. 78–86 (2015)
18. Singh, G., Nelson, A., Robucci, R., Patel, C., Banerjee, N.: Inviz : Low-power Personalized Gesture Recognition using Wearable Textile Capacitive Sensor Arrays
19. Nelson, A., Singh, G., Robucci, R., Patel, C., Banerjee, N.: Adaptive and Personalized Gesture Recognition using Textile Capacitive Sensor Arrays. IEEE Trans. Multi-Scale Comput. Syst., 1–1 (2015)
20. Braun, A., Wichert, R., Kuijper, A., Fellner, D.W.: Capacitive proximity sensing in smart environments. J. Ambient Intell. Smart Environ. **7**, 1–28 (2015)
21. Grosse-Puppendahl, T., Berghoefer, Y., Braun, A., Wimmer, R., Kuijper, A.: OpenCapSense: A rapid prototyping toolkit for pervasive interaction using capacitive sensing. In: 2013 IEEE International Conference on Pervasive Computing and Communications, PerCom 2013, pp. 152–159 (2013)
22. Tapia, E.M.: Activity Recognition in the Home Setting Using Simple and Ubiquitous Sensors. Technology. LNCS, vol. 3001, pp. 158–175 (2004)
23. Lara, O.D., Labrador, M.A.: A survey on human activity recognition using wearable sensors. Commun. Surv. Tutorials, IEEE. **15**, 1192–1209 (2013)
24. Mitchell, H.B.: Multi-sensor Data Fusion. Springer (2007)
25. Weichert, F., Bachmann, D., Rudak, B., Fisseler, D.: Analysis of the accuracy and robustness of the leap motion controller. Sensors **13**, 6380–6393 (2013)
26. Shotton, J., Fitzgibbon, A., Cook, M., Sharp, T., Finocchio, M., Moore, R., Kipman, A., Blake, A.: Real-time human pose recognition in parts from single depth images. Commun. ACM **56**, 116–124 (2013)
27. Ono, M., Shizuki, B., Tanaka, J.: Touch & activate: adding interactivity to existing objects using active acoustic sensing. In: Proceedings UIST, pp. 31–40 (2013)
28. Rekimoto, J.: SmartSkin: an infrastructure for freehand manipulation on interactive surfaces. In: Proceedings CHI, pp. 113–120 (2002)
29. Wilson, A., Pham, H.: Pointing in intelligent environments with the worldcursor. In: Proceedings International Conference on Human Computer Interaction (2003)

30. Majewski, M., Braun, A., Marinc, A., Kuijper, A.: Providing visual support for selecting reactive elements in intelligent environments. Trans. Comput. Sci. XVIII. **7848**, 248–263 (2013)
31. Wilson, A., Benko, H., Izadi, S., Hilliges, O.: Steerable augmented reality with the beamatron. Proc, UIST (2012)
32. Rus, S., Grosse-Puppendahl, T., Kuijper, A.: Recognition of Bed Postures Using Mutual Capacitance Sensing. In: Ambient Intelligence, pp. 51–66. Springer (2014)
33. Braun, A., Frank, S., Majewski, M., Wang, X.: CapSeat: capacitive proximity sensing for automotive activity recognition. In: Proceedings of the 7th International Conference on Automotive User Interfaces and Interactive Vehicular Applications, pp. 225–232. ACM, New York, NY, USA (2015)
34. Braun, A., Heggen, H., Wichert, R.: CapFloor—A Flexible capacitive indoor localization system. In: Proceedings Evaluating AAL Systems Through Competitive Benchmarking. Indoor Localization and Tracking, pp. 26–35 (2012)
35. Braun, A., Zander-Walz, S., Krepp, S., Rus, S., Wichert, R., Kuijper, A.: CapTap—Combining Capacitive Gesture Recognition and Knock Detection. In Proceedings of the 3rd International Workshop on Sensor-based Activity Recognition and Interaction iWOAR (2016)

LivingCare—An Autonomously Learning, Human Centered Home Automation System: Collection and Preliminary Analysis of a Large Dataset of Real Living Situations

Ralf Eckert, Sebastian Müller, Sebastian Glende, Alexander Gerka, Andreas Hein and Ralph Welge

Abstract Within the scope of LivingCare, a BMBF funded research project, a real senior residence was equipped with a large amount of home automation sensors. More than sixty sensors and actuators were installed in this apartment. All actions performed by humans like switching light on or off, setting the temperature and the usage of electric devices like TVs will be recorded. This data is collected over a period of 18 months. Thus, one of the largest mobility and characteristics datasets based on home automation sensors will be acquired. This data will be the foundation for developing autonomously learning algorithms. During the second project phase these algorithms will start to control functions of the home automation system. The project's objective is to develop an autonomously learning home automation system that automatically adapts to the residents' behavior. The system

R. Eckert (✉) · A. Gerka
OFFIS-Institute for Information Technology, Oldenburg, Germany
e-mail: eckert@offis.de

A. Gerka
e-mail: gerka@offis.de

S. Müller · R. Welge
Leuphana University of Lüneburg, Lüneburg, Germany
e-mail: sebastian.mueller@leuphana.de

R. Welge
e-mail: welge@leuphana.de

S. Glende
YOUSE GmbH, Berlin, Germany
e-mail: sebastian.glende@youse.de

A. Hein
Carl von Ossietzky University Oldenburg, Oldenburg, Germany
e-mail: Andreas.Hein@informatik.uni-oldenburg.de

R. Wichert and B. Mand (eds.), *Ambient Assisted Living*,
Advanced Technologies and Societal Change,
DOI 10.1007/978-3-319-52322-4_4

will be able to grow with the users' needs. With all the possible data collected it will be able to support daily actions, recognize behavior changes over time and will be able to call help in emergency situations.

1 Overview

The research work described in this paper is taking place in the research project LivingCare. The LivingCare research and development project is funded by the German Federal Ministry of Education and Research (BMBF) within the Framework concept "Adaptive Learning Systems" (grant 16SV7206).

In this project, the consortium is working on a human centered home automation system to support elderly people in their own homes and to assure a longer and self-sufficient life. To achieve these objectives several criteria will be addressed:

- The utilized home automation has to be cost-efficient and has to be easily installed in existing housing structures like houses and apartments.
- A novel sensor for detecting a dependable number of persons in a room will be developed.
- The home automation system will not be driven by inflexible rules but by autonomous learning algorithms.

Particular attention during the software development will go to the design of the autonomous learning algorithms. The final system will be able to perform specific actions, such as turning on and off lights or electrical devices and managing the indoor temperature, autonomously. To avoid confusing the user witch possibly obscure actions performed by the home automation system, it is necessary to test and evaluate those algorithms with real data. In the following, beside a general project overview, we will describe how the required data for the evaluating process was recorded, what was analyzed and what information was detectable. The focus lies primarily on movement-behavior and behavior patterns.

2 State of the Art

The state of the art of LivingCare contains several disciplines described next. We will briefly touch each of the subjects home automation, mobility and behavior modeling and analysis, data mining and stochastic processes.

2.1 Home Automation

Home automation (HA) focuses, as a part of building automation systems, especially on the conditions of private homes and the special demands of its inhabitants.

HA sensors and actors are used primarily for illumination, shading, heating, ventilation and air conditioning. Other applications are alarm systems and energy saving.

Existing systems could be roughly distinguished in cable based and wireless solutions. In cable based solutions, the communication works over a bus. Known standard bus base systems are, for example, KNX, LonWorks and BACnet. In the case of wireless systems, no construction measures have to be performed. That's why it is much easier to use these systems in existing buildings. Known radio standards are ZigBee, Z-Wave, EnOcean, KNX RF and other manufacturer specific solutions.

During commissioning, the system integrator and the client define rules for the HA system. These rules can be altered later, but not without the system integrator. This is very inflexible. Systems that learn from user input would be more flexible and accepted by users.

2.2 Mobility Analysis and Behavior Modeling

Personal mobility is an important condition for a self-reliant and socially integrated life [1], especially the ability to move with or without technical aids from one place to another. The change of mobility, especially in old age, is correlated with health deterioration. Common tests for elderly people are gait and balance tests. In many studies, technical systems were introduced based on ambient or wearable sensors to objectively determine a person's mobility [2]. Research done by OFFIS in the projects PAGE [3] and GAL [4] have shown that the usage of ambient sensors leads to saving of time in clinical applications and that it is much easier to install such systems at home for preventive and rehabilitative applications. Home automation can be used for those applications due its easy installment, as has been shown previously [5–9].

2.3 Statistical Data Mining and Stochastic Processes

Methods of statistical data analysis are used to gather knowledge of process data. Analysis of sensor data allows to classify situations, which in turn is a starting point for context sensitive home automation services. The process of knowledge discovery can be separated into the phases of providing data sources, target definitions, data selection, data cleansing, data reduction and transformation, process model selection, data analysis and knowledge interpretation. For preprocessing, the following filter techniques are known: polynomic based approaches like Savitzky-Golay and outlier tests by Nalimov, Walsh or Grubbs. For multivariate datasets, Andrews-Curves, M-estimators or Stahel-Donoho-Outlyingness are suitable to find artefacts and abnormal process factors. Following the data analysis, we can classify process data with the assistance of methods like k-means, Maximum Margin Clustering or EM-Clustering. To determine process parameters (e.g. Gaussian Mixture Models) for stochastic models, methods of descriptive and explorative statistics are used. The methods main objective is the identification of features vectors that can be associated with states of probabilistic automata (SD HMM- State Duration HMM). Subsequently, Hidden Markov Models (HMM) and Relational State Descriptions for modeling spatial and temporal states can be used. Posture- and gesture recognition based on HMMs [10] and the recognition of human activity within the Ambient Assisted Living (AAL) domain with linked HMMs was subject of discussion previously [14, 15]. Methodical approaches are pursued by Moghaddam and Piccardi [11]. Shiping et al. address the training of HMMs of second order with the help of multiple observation sequences [12]. State duration approaches are proposed by Levinson [13].

3 Own Approach

Home automation, as used today, is an inflexible entity. Once it is programmed, it performs always in the same manner. State A always leads to state B. The advantage of a deterministic system is that its behavior is predictable and errors are easy to recognize. In LivingCare, the system will adapt autonomously to the users' needs. It will grow with the rising and changing needs of its user to maintain the possibility to live an independent, secure life as long as possible. This is not possible with a system depending on inflexible rules. That's why the home automation system in LivingCare will be driven by autonomous learning algorithms. This provides some advantages:

- Behavior changes will be recognized in an earlier state. So it will be possible to react in time when, e.g. the person stops preparing meals or using the shower.
- The system will be able to adapt to new behavior and support the inhabitant as much as possible.
- For a commercial product, service will be cheaper because it is not required to call a technician for every small system adaption.
- It is possible to install all components at once, but modules and functions can be activated and deactivated according to the users' needs.

An autonomously acting system may cause plenty of risks. If the system does not work as expected it could cause confusion and possibly even harm to the user. That must be avoided at any circumstances.

The autonomously learning algorithm are based on reinforcement learning methods. The algorithm will monitor the inhabitants' behavior for a certain time to learn their habits. After the learning phase, some simple actions, such as switching on lights or adjusting temperature, will be performed. Each performed corrective action by the user will be recognized as negative feedback. The algorithm will adjust its sets of states and actions behave differently next time. Nevertheless, the system will only be able to act in tight boundaries. The temperature for example will only be adjustable in a specified range and the algorithm is only able to turn lights on but not off to avoid injuries.

3.1 Scenarios

To keep the project manageable, it was essential to identify some important scenarios that would provide a real benefit for elderly people in their homes. We identified these scenarios:

- **Needs-oriented illumination:** Illumination will be set according to the users' needs. These needs will be derived from the user habits and external factors. The system learns to link both information sources together and will use this information to generate an optimal illumination scenario.
- **Needs-oriented indoor climate:** the home automation system provides everything needed to control room climate. In combination with user models, the algorithm will be able to provide the suitable room climate for every situation. Particular persons will be able to control the temperature via Internet.
- **Movement sensitive alarm:** The system will recognize absence and sleeping times. If one of both states is identified, a burglar alarm is automatically activated.
- **Safe Mode:** If the user leaves the home for longer time, the system will shut down all devices the could cause fire or will waste energy, like stoves or electric irons.
- **Monitoring of adults in need of care:** The used sensors allow to monitor resident behavior. The occurrence of abnormal behavior, like distinctly long presence in the bathroom, could be identified and lead to an emergency call. Slow changes in behavior would be identified and adapted to. If the user spends less time in the kitchen day by day, it could implicate that the person doesn't cook meals anymore and an appropriate information could be delivered to individuals involved in this resident's care.
- **Intelligent pill-cabinet:** A sensor, equipped to the cabinet door, will recognize if the door was opened or not. A reminder will be given to the user and if that does not lead the user to take the pills, an authorized person will be contacted.
- **Visual and acoustic telephone bell:** Telephone calls will be indicated through light and sound.
- **Mailbox sensor:** a sensor will be placed in the mail box. The user will be informed if mail was delivered via light, sound or a tablet app.
- **Remote maintainable AAL-components:** the system will able to self-monitor. In case of problems it should be possible to fix the system via remote maintenance.

4 Realization

To secure failure free function, the algorithm will be verified against real data. Dysfunction in the real application must be minimized. Behavior data, based on home automation systems, were not available in the amount needed for proper testing. So the first task was to set up a home automation environment in a real apartment inhabited by real people. The apartment was equipped with all available conventional kinds of home automation sensors provided by EQ 3 s *homematic* system. Data from these sensors will be gathered for about 18 months. This will result in one of the largest datasets of that kind gathered in the scope of an AAL research project. The data will be analyzed and used as a learning base for the reinforcement algorithm.

4.1 Test Persons/Subjects

It is not easy to find volunteers in the designated target group. Elderly people often refuse modern technology or are afraid not to understand what happens to them or that they will be overstrained. The Deutsches Rotes Kreuz (DRK) is one of the project members. The DRK runs a retirement complex in Oldenburg, Germany, with 68 residential units. The different types of apartments range from small one room apartments to bigger penthouse apartments. The DRK was asked to find volunteers out of these apartments for the 18-month field trial. No special requirements had to be fulfilled. It was more important to find anyone who was willing to help. The search was successful: A married couple living in one of the penthouses was happy to help.

Test person 1 is marked by the following characteristics:

- 75 years old female
- very hale and active
- good mobility
- visits sports groups regularly
- takes dog for walks daily

Test person 2 is marked by the following characteristics:

- 82 years old male
- very restricted mobility
- high risk of falling
- mentally fit
- not as active as test person 1
- light smoker

Housework like cleaning up, laundry, ironing etc. is done by test person 1. Once a week, test person 1 is assisted by a cleaning lady. Both test persons are technically affine and use smartphones and personal computers.

The project background was explained in detail to both. Before installing the sensors, both were asked for their approval and the ethics committee of the University of Oldenburg was asked for approval. Both test persons know they can always terminate their participation without giving reasons. At any time, they will be explained in detail about what data were collected in an easy to understand way.

4.2 The Penthouse

The floor plan of the penthouse is shown in Fig. 1. The penthouse provides about 80 square meters of living space, divided into two bedrooms, a kitchen, a large living room, a shower room with toilet, a small corridor to separate the bedrooms and bathroom from the rest of the apartment, and a longer corridor with an integrated kitchenette. Furthermore, there is a sheltered Loggia and a roof terrace accessible from both bedrooms and the living room. The penthouse is situated on the 5th floor.

In the field of accessibility, the following features of the housing arrangement should be highlighted:

- Elevator up to the 4th floor
- Stair lift from 4th to 5th floor
- Ground level walk-in shower
- Support rails in shower and next to toilet
- To access the loggia, a 25 cm tall ring foundation must be climbed. Since this implies a high risk of falling, a support rail is mounted next to the loggia door.
- Nurse call in the livingroom, the longer corridor and both bedrooms.

4.3 Home Automation

Equipping existing buildings with home automation was a very cost-intensive effort just a few years ago. Those systems normally were linked by cable (Bus), or a special kind of cable installment (star topography wiring) was required. That's why normally only new buildings were equipped with home automation. The installation still wasn't cheap: To equip a single family house with home automation, expenses of more than 10.000 € were common. Nowadays, much cheaper retrofit solutions are on the market. These solutions are often radio controlled and no additional cables are required. Actuators and sensors often replace common elements such as

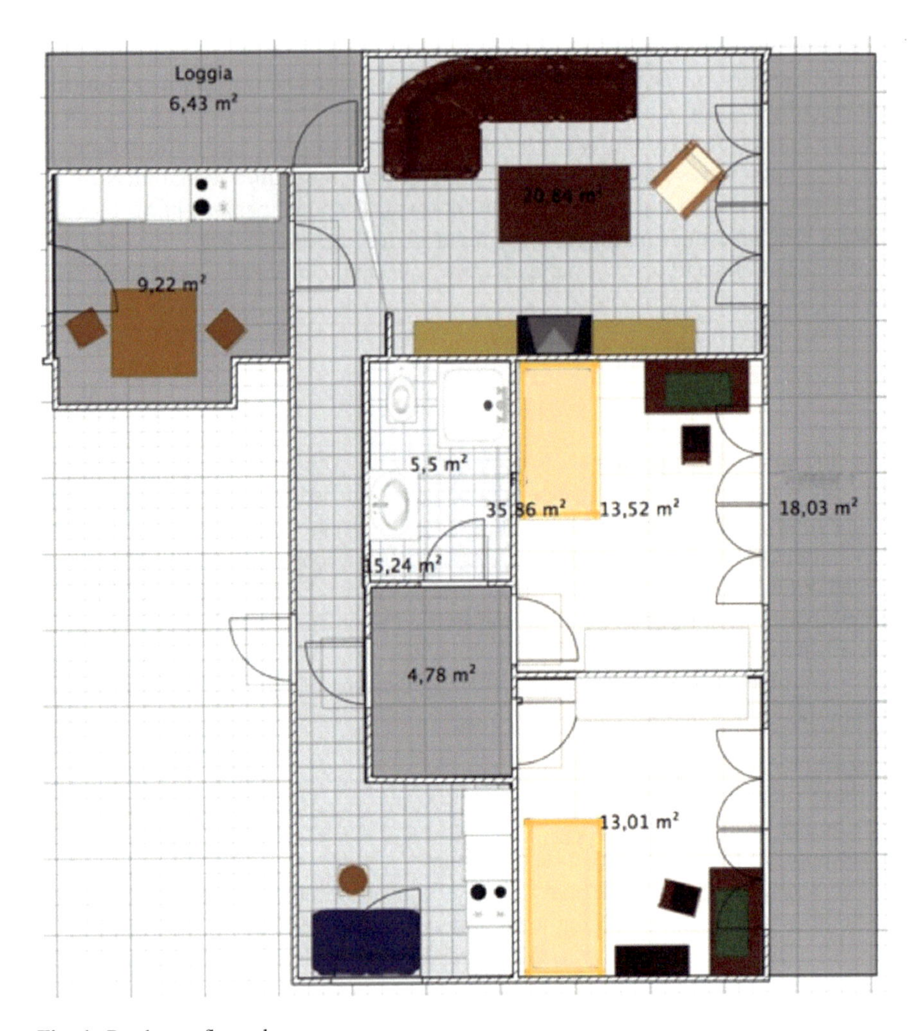

Fig. 1 Penthouse floor plan

light switches or standard sockets. Switching actuators are so flat they can be used underneath regular components or in junction boxes. Remote controlled home automation has one drawback: many components are battery-powered. This causes a considerable amount of waste over the years. n selection and functionality, the novel systems are as good as the established systems like e.g. KNX/EIB. Due to the described properties, home automation becomes more and more popular, especially in existing buildings and in private homes. Even new buildings are being equipped with the cheaper remote controlled based systems. Instead of costs in the 10.000 € range, a fully equipped private home will cause costs around 2 to 3.000 €.

4.4 Sensors

During the field trial from June 2015 to Dec 2016, all kinds of data is being collected by a home automation system to generate movement profiles and detect behavior patterns. A home automation system developed by EQ 3, *homematic*, is used in this project. To achieve these goals, it is important to specify the used home automation components:

- **Door and window contacts:** With these sensors it is possible to detect when windows or doors are opened or closed. In combination with motion sensors, it is possible to extract movement directions and speed.
- **Motion sensor:** When motion sensors are mounted correctly they allow to detect presence in rooms and motion paths from one room to another, as well as movement speed. Furthermore, an additional brightness sensor is built into the sensor.
- **Switch actuators:** Used as a replacement for conventional light switches, it is possible to detect when the inhabitants turn lights on and off. In combination with the motion sensors' brightness sensor it is possible to draw conclusions about lighting conditions, too.
- **Roller shutter actuator:** This sensor allows to detect the usage of roller shutters and their positioning.
- **Electrical radiator thermostat:** By the usage of this sensor/actuator it is possible to detect the actual and selected temperature.
- **Wall-mounted thermostat:** The additional benefit of the radiator thermostat is the build in hygrometer and it is possible to regulate more than one radiator at once.
- **Switchable sockets with power meter:** This sensor is connected to electrical devices of daily use like television sets, microwave oven, floor lamps, computers or similar.

The persistence of all data flows is handled by a base station. The entire system is not connected to the internet to ensure data safety with highest certainty. Every 14 days an authorized person visits the subjects to pick up the data by hand. The correct system functionality is checked at the same time as well as the subjects' satisfaction with the system.

4.5 Installation

Altogether, a total number of 64 sensors and one base station were installed in the penthouse. The support of an external electrician was required. The whole process lasted about one day. Another day was used to set up and test the system.

The complete installation is running since July 2015. Minor problems were fixed expeditiously. In the beginning, two wall-mounted thermostats fell off the wall, because of flimsy adhesive. Another problem was a corridor intermediate switch circuit that doesn't work properly.

In detail the following sensors were installed:

- 10 movement sensors,
- 7 electrical radiator thermostats,
- 3 wall-mounted thermostats,
- 11 door and window contacts,
- 12 switchable sockets with power meter,
- 14 switch actuators,
- 6 roller shutter actuators,
- 1 junction box mounted power meter.

The only holes drilled were for 10 movement sensors and the 3 wall-mounted thermostats. The holes are very small and, if required, easy to seal, so no visible damage would remain after removal.

This experience shows that an installation of that kind could be accomplished without major problems. Later commercial systems should cause no mentionable effort.

5 Results

The system now is active for about 4 months without any negative incident and is collecting data without interruption. During that period, more than 55.000 sensor events were recorded weekly. The dataset currently consists of more than half a million records and is steadily growing. The next step is to analyze and structure this data. Our first objective is to develop a data mining algorithm which, based on the recorded data, is able to detect sensor activation sequences that represent typical inhabitant behavior through different sub-steps. These detected sequences are used as input for the autonomous learning algorithm. In the final project stages, the algorithm should be able to recognize typical inhabitant behavior, to detect atypical behavior and detect slow behavior changes over time. Based on these results, the algorithm generates home automation rules that will support the inhabitants' daily life.

Preliminary projects showed that normal behavior and deviations of that behavior could be recognized by observation of activation sequences and duration of sensor events. But without knowledge about the floor plan, sensor installation locations and other prior knowledge this method may not work correctly. Different to prior approaches, this algorithm should work with as little prior knowledge as possible and standard algorithms should be used as far as possible.

5.1 Method

First, an explorative data analysis was undertaken. The objective was to evaluate the information content and to gain a first impression about "recognizable" behavior patterns in the sensor data and to find failures in sensor location and to give optimization proposals. Another approach was to determine the minimum prior knowledge required to recognize behavior patterns.

5.2 Association Analysis

Association analysis was used to determine what regular combinations of sensors and sensor data could be found. The association analysis rules describe the correlation among mutually occurring events. The purpose lays in determining elements of actions that imply the occurrence of other elements of the same action.

In the end, association analysis does not provide satisfying results. It was not possible to recognize related elements reliably. The frequent occurrence of certain sensor events in relation to its total appearance does not say anything about its significance. Moreover, the sensor event order differs too much to determine sequences that are similar, but not identical.

5.3 General Activity

A simple form of behavior at home is to move from one room to another or to stay in a room for a while. To determine movement and mobility, movement sensors were installed so that every 4 s an event was triggered if movement was detected in the field of view.

To be able to make that kind of activity visible, general activity was modeled so that the frequency of sensor activations for every hour over a day were accumulated. The more frequently a specific sensor was activated over the day, the more activity took place in that room.

At first the sensor events were filtered based on sensor types. Only events triggered by movement sensors were analyzed. Afterwards every event of every sensor every hour per day were counted and the mean value for each day was calculated. The results are shown in Fig. 2 in form of a heat map.

As a result, it can be specified that without any prior knowledge, only through movement sensor data analysis, it is possible to make rough statements about the daily activity level of inhabitants. The diagram shows, for example, that the kitchen is used more frequently around lunch time and supper time. *KBewegung* identifies the kitchen movement sensor. *SMBewegung* describes the movement sensor in the

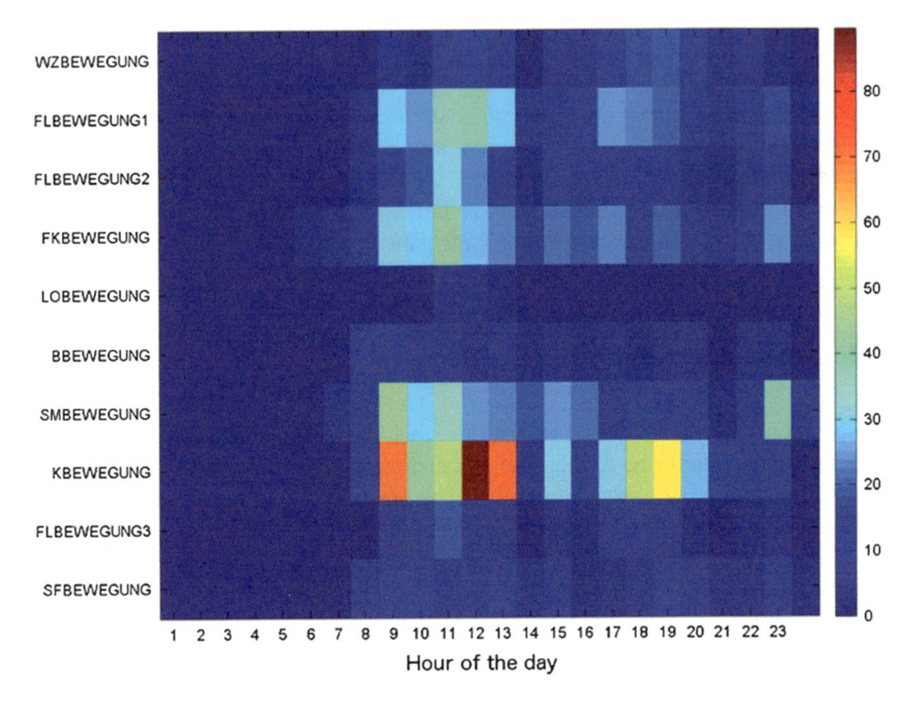

Fig. 2 Accumulated movement sensor events per hour over one day

bedroom of test person 2. It is easy to see when the person is going to bed and when the person gets up in the morning.

5.4 *Activity Detection Through Analysis of Electrical Parameters*

In a previous project called "AmbiACT", it was demonstrated that it is possible to recreate daily routines, general behavior and abnormal behavior through the use of electrical devices. Therefore, switchable sockets with an integrated power meter were installed at the test apartment. This time, instead of movement sensor events, as in the "general activity" section, the events triggered by the power meters were analyzed. The objective was to show if the used energy per measuring point per hour over a day allows to detect general activity and a daily routine.

Figure 3 shows the evaluation results. For example, the red graph *WZMESSTV* shows when and how long the television in the living room was used. The television is mostly used in the evening hours between 5 p.m. and 10 p.m. The result shows that it is quite easy to conclude that observing energy consumption of specific devices makes it possible to determine daily routines. It must be noted that prior knowledge is needed in this case: Without the knowledge what sensor is connected to witch device it is not possible to recognize specific activities.

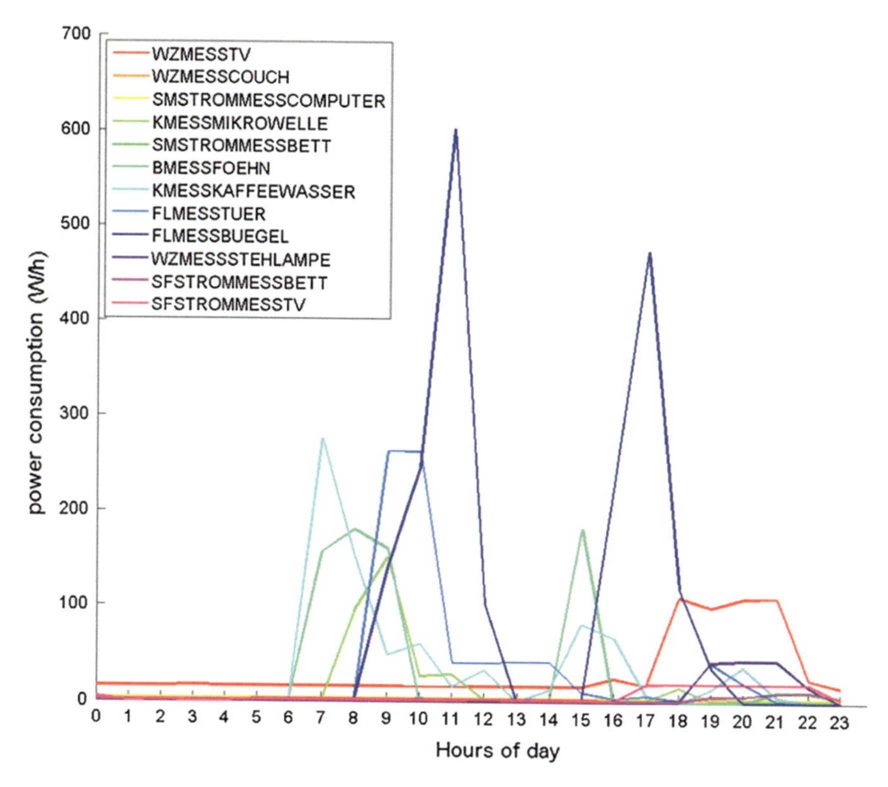

Fig. 3 Activity detection through analysis of electrical parameters

5.5 *Determining Occupation Times*

A lot of accidents at home involving seniors happens in the bathroom. The floor is often slippery and when someone falls he or she falls hard. That's why it is important to determine occupation times, especially for the bathroom. To recognize unusual long occupation times is very safety relevant for the inhabitants. Thus, another approach is to examine the significance of movement sensors and door-contacts to determine occupation times for the bathroom. Door-contacts alone do not help to determine occupation times because the door is not always closed when the room is in use. Again, movement sensors are very useful for this scenario. Every 4 s a movement is detected in the field of view an event is triggered. To calculate the occupation time, all events occurring every 4 s have to be accumulated. The result is the time a room was occupied without interruption.

Figure 4 shows the median occupation time for the bathroom. An error-bar graph is used for visualization. It shows the median occupation on the y-axis and the standard deviation on the x-axis according to the hour of the day. The longest occupation time for the bathroom is obviously in the morning hours between 8 a.m. and 10 a.m. The high standard deviation may imply that the shower is not used every day.

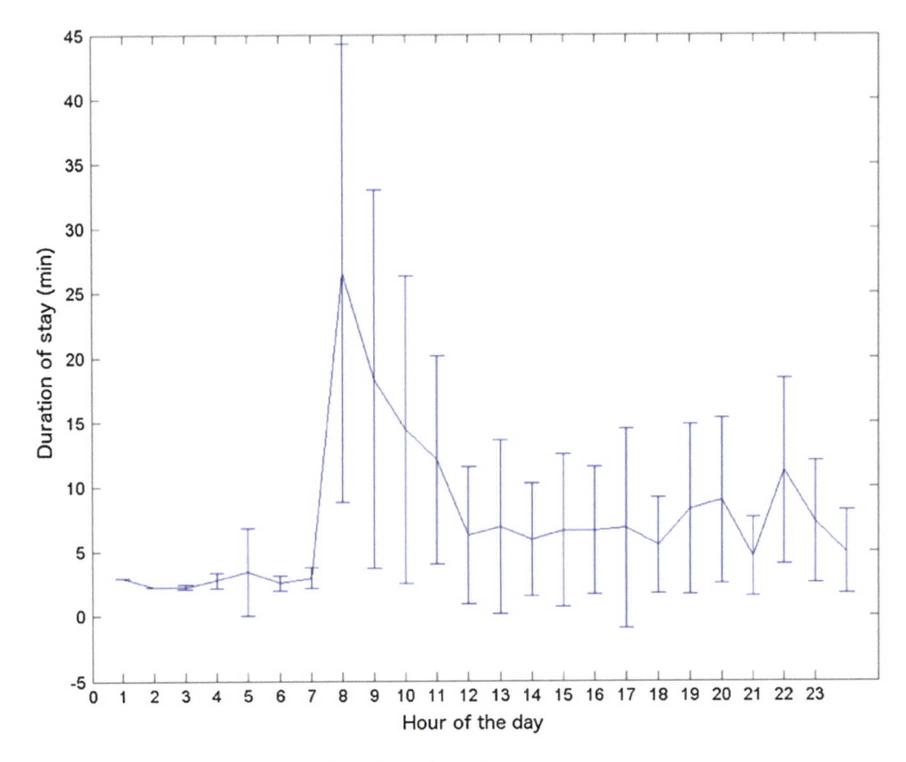

Fig. 4 Determining occupation times through motion sensors

The evaluation showed that it is possible, with the usage of movement sensors, to determine the habitual occupation time of a specific room. If this method would be used as an alarm trigger for too long room occupation, in this example an alarm should be triggered in the mornings after an occupation of more than 60 min and 30 min for the rest of the day with a low risk of false alarm. The only required prior knowledge is the name of the movement sensor to determine the correct room. The additional analysis of door-contacts does not improve the results.

5.6 Estimation of Walking Speed and Recognition of Persons with Movement Sensor Data

Self-selected gait speed is an important indicator for current health problems and an indicator of future problems. Although gait speed generally decreases with age, sudden changes and gait speeds below 0.5 m/s are strong indicators of health issues.

To detect gait speed using home automation sensors, we choose two motion sensors that have a long and ideally straight distance between them, and measure the time between sensor events: the area of the hallway between the kitchen and

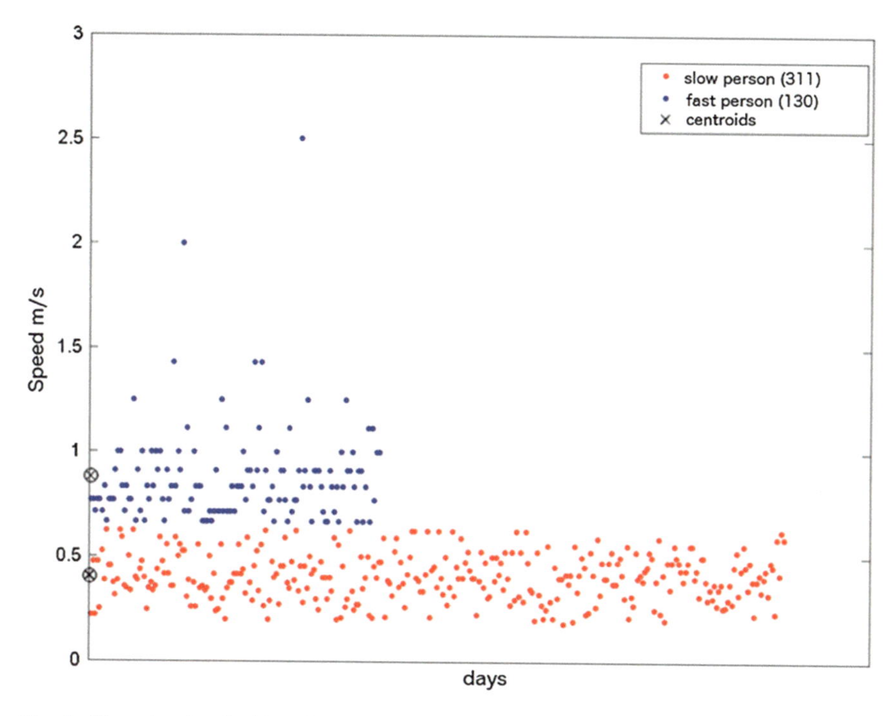

Fig. 5 Clustering for the identification of different movement profiles

kitchenette is an ideal place for this measurement. We can find the distance between the motion sensors using the blueprint of the flat (Fig. 1). Although not all measurements between the two sensors will be precise, as subjects may stop in between, turn around or engage in some activity, using several measurements over a longer period of time gives us an estimate of the actual gait speed.

Although the sensor data does not tell us which subject walked between the sensors, we can apply a cluster analysis on the speed of each measurement. Figure 5 shows that, at least for part of the time, there is a constant occurrence of slow (0.38 m/s) and fast (0.78 m/s) movements. While we cannot be certain whether both subjects are clearly separable using gait speed alone, the findings coincide with our knowledge of the subjects' mobility.

5.7 *Detection of Behavior Patterns*

Daily activities contain countless reoccurring short and long sequences, which in part can be found in the collected data. Part of the project is to test the data for discoverable rituals or sequences that, at least in part, can be automated. The current subjects informed us of one particular daily ritual: One of the subjects watches TV until approximately 23 h. During this time, the floor lamp in the living room is lit. When the subject goes to bed, he must first turn on the light in the hallway or kitchen, then returns to the living room to turn off TV and light. From the hallway, the subject goes to the loggia to smoke a cigarette. After that, he goes to the bedroom to turn on the light, and then back to the hallway to turn off the light there.

In terms of automated recognition of such patterns, it is important to note that neither the order of the events nor the exact composition of events is precisely defined. For instance, the other subject may turn on the light in the hallway, or the light might still be on, thus no motion in the hallway would be detected.

To test whether this pattern can be found in the data, we chose data of the evening times of 35 consecutive days. First, all data from sensors and actuators not involved in this pattern is removed. Second, using information on the placement of sensors, sequences of activity of neighboring sensors are detected.

Of the 35 days, the sequence was found on 28 days. The time of occurrence of the sequence is shown in Fig. 6. The time of occurrence falls between 21:47 and 22:46 each day.

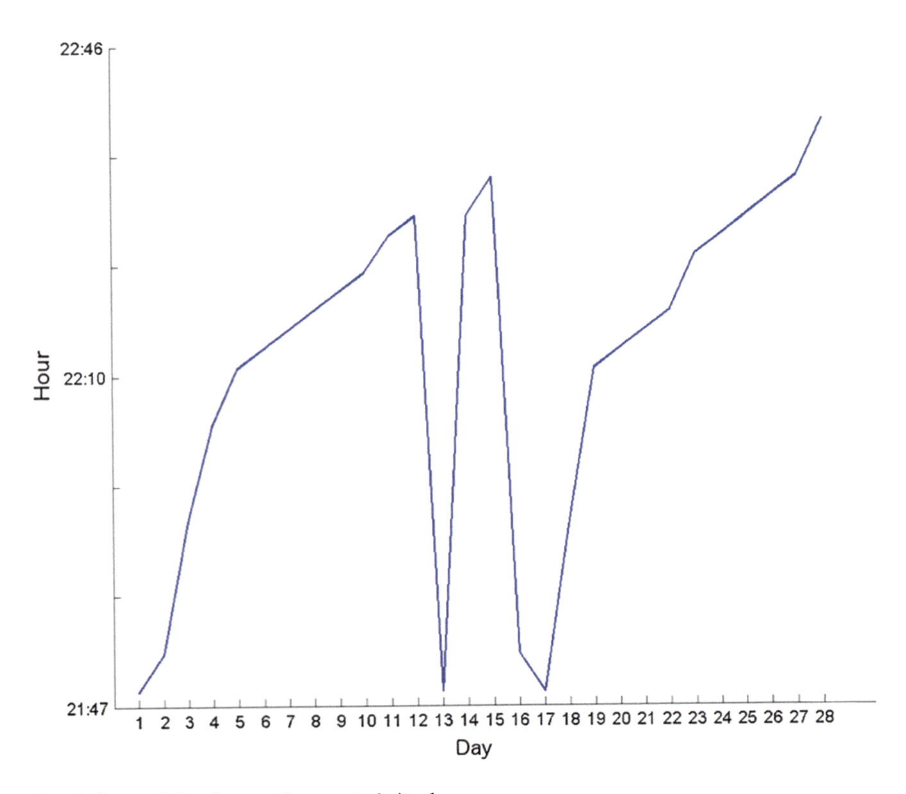

Fig. 6 Determining frequently repeated rituals

6 Discussion

Although the project is still in an early stage, and further data analysis is necessary, it could be shown that certain behavior patterns, such as gait speed and presence by room, can be derived from the data. Furthermore, it is possible to find complex recurring behavior patterns with minimal effort.

The next step in the system development will be the deduction of automatable actions, i.e. finding patterns of behavior that can be executed by the LivingCare system automatically. For example, in the process of going to bed, the system might turn of lights and devices as subjects are leaving the room. To gain broad acceptance in a mass market, the detection of such patterns and actions must work reliably, thus particular effort must be made to implement these features error-free.

Lastly, the reinforcement algorithms will help to understand if certain automation steps are desired. Besides the obvious technical difficulties, home automation still requires the users' acceptance: while automation of rituals may be perceived as facilitation, some behavior patterns may be personally valuable to the user, or be supportive in maintaining physical and mental health.

To implement all aspects of the system, much work still needs to be done in regard to hardware and software development.

References

1. Deutsches Institut für medizinische Dokumentation und Information (DIMDI): Internationale Klassifikation der Funktionsfähigkeit, Behinderung und Gesundheit (2005)
2. Scanaill, C.N., Carew, S., Barralonm, P., et al.: A review of approaches to mobility telemonitoring of the elderly in their living environment. Ann Biomed Eng **34**(4), 547–563 (2006)
3. Frenken, T., Lipprandt, M., Brell, M., Wegel, S., et al.: Novel approach to unsupervised mobility assessment tests: field trial for aTUG. In: Proceedings of the 6th Int Pervasive Computing Technologies for Healthcare (PervasiveHealth) Conf, pp. 131–138 (2012)
4. Hein, A., Winkelbach, S., Martens, B., et al.: Monitoring systems for the support of home care. Inf. Health Soc. Care **35**(3–4), 157–176 (2010)
5. Helmer, A., Lipprandt, M., Frenken, T., Eichelberg, M., Hein, A.: 3DLC: a comprehensive mode for personal health records supporting new types of medical applications. J. Healthcare Eng. **2**(3), 321–336 (2011)
6. Steen, E.E., Frenken, T., Eichelberg, M., Frenken, M., Hein, A.: Modeling individual healthy behavior using home automation sensor data: Results from a field trial. J. Ambient Intell. Smart Environ. (JAISE) **5**(5), 503–523 (2013)
7. Hadidim T., Noury, N.: A predictive analysis of the night-day activities level of older patient in a health smart home. In: Proceedings of the 7th International Conference on Smart Homes and Health Telematics: Ambient Assistive Health and Wellness Management in the Heart of the City, pp. 290–293 (2009)
8. Floeck, M., Litz, L.: Activity- and inactivity-based approaches to analyze an assisted living environment. In: Proceedings of the 2008 2nd International Conference on Emerging Security Information, Systems and Technologies, pp. 311–316. IEEE Computer Society (2008)

9. Skubic, M., Guevara, R.D., Rantz, M.: Testing classifiers for embedded health assessment. In: Proceedings of the 10th International Smart Homes and Health Telematics Conference on Impact Analysis of Solutions for Chronic Disease Prevention and Management, ICOST'12, pp. 198–205 (2012)
10. Liu, N., Lovell, B.C.: Gesture classification using Hidden Markov Models and Viterbi path counting. In: Proceedings of the 7th Digital Image Computing: Techniques and Applications (2009)
11. Moghaddam, Z., Piccardi, M.: Deterministic initialization of hidden markov models for human action recognition. In: Proceedings of IEEE Digital Image Computing: Techniques and Applications (2009)
12. Shiping, D., Tao, C., Xianyin, Z., Jian, W., Yuming, W.: Training second-order Hidden Markov Models with multiple observation sequences. In: Proceedings of the IEEE International Forum on Computer Science-Technology and Applications (2009)
13. Levinson, S.E.: Continuously variable duration hidden Markov models for automatic speech recognition. In: Computer Speech and Language, pp. 29–45 (1986)
14. Busch, B.H., Kujath, A., Witthöft, H., Welge, R.: Preventive emergency detection based on the probabilistic evaluation of distributed, embedded sensor networks. Ambient Assisted Living, 4. AAL-Kongress 2011, Berlin, Germany, 25–26 Jan 2011. Springer (2011)
15. Busch, B.H., Welge, R.: Domain specific services for continuous diagnoses in the context of ambient assisted living-AAL. In: Proceedings of the International Conference on Data Mining (DMIN'11), 2011

New Approaches for Localization and Activity Sensing in Smart Environments

Florian Kirchbuchner, Biying Fu, Andreas Braun and Julian von Wilmsdorff

Abstract Smart environments need to be able to fulfill the wishes of its occupants unobtrusively. To achieve this goal, it has to be guaranteed that the current state environment is perceived at all times. One of the most important aspects is to find the current position of the inhabitants and to perceive how they move in this environment. Numerous technologies enable such supervision. Particularly challenging are marker-free systems that are also privacy-preserving. In this paper, we present two such systems for localizing inhabitants in a Smart Environment using—electrical potential sensing and ultrasonic Doppler sensing. We present methods that infer location and track the user, based on the acquired sensor data. Finally, we discuss the advantages and challenges of these sensing technologies and provide an overview of future research directions.

Keywords Smart environment · Electric potential sensing · Ultrasonic sensing · Data mining · Indoor localization

1 Introduction

Indoor localization has been an important area of research in ubiquitous computing and associated research areas for several decades. The visual detection and tracking of an object are simple tasks for humans. Having knowledge of the subject mat-

---

F. Kirchbuchner · B. Fu · A. Braun · J. von Wilmsdorff (✉)
Fraunhofer Institute for Computer Graphics Research IGD,
Fraunhoferstr. 5, 64283 Darmstadt, Germany
e-mail: julian.von.wilmsdorff@igd.fraunhofer.de

F. Kirchbuchner
e-mail: florian.kirchbuchner@igd.fraunhofer.de

B. Fu
e-mail: biying.fu@igd.fraunhofer.de

A. Braun
e-mail: andreas.braun@igd.fraunhofer.de

R. Wichert and B. Mand (eds.), *Ambient Assisted Living*,
Advanced Technologies and Societal Change,
DOI 10.1007/978-3-319-52322-4_5

ter and the situational context, one can perceive objects, detect motion, and make predictions. Occlusions, intersections, or rotations often do not cause any problems. Accordingly, many indoor localization systems are based on one or more cameras, thus replicating the human visual system [16]. However, there are considerable challenges and limitations to using this approach, since the resolution of cameras is typically much lower, and the system is far less accurate. Moreover, the human brain has been trained for many years on object detection tasks. However, instead of imitating the human eye, it also makes sense to investigate alternative approaches. Reliable outdoor navigation techniques have been used for centuries, and they are continuously refined. Nowadays, many users already apply smartphone-based localization systems, employing GPS sensors to retrieve their current location. However, GPS-based systems cannot be used inside a building due to the limitations of blocked communication lines, which are caused by the global positioning of satellites and the strong attenuation of microwaves by materials used in walls and roofs [23]. In emergency situations, such as fire or medical emergencies, it would be beneficial to know the current position of the inhabitants within a building. Additionally, in elderly homes or assisted-living facilities, such an indoor localization system can provide additional safety and comfort functions and thus support the users if necessary. Localization can also be used for behavioral analysis, including the detection of irregularities in the routines of people and with respect to their everyday activities.

The existing techniques for indoor localization can be roughly divided into two categories. Basically, we can distinguish between marker-based and marker-free tracking methods for indoor localization [20]. Wearable and marker-based systems are very common but have one major disadvantage: people have to wear a tag to actively use the systems. Particularly for older adults, it was shown in [4] that markers are often deliberately not worn or simply forgotten. The same study found out that wearable devices are also often rejected by potential users for fear of being stigmatized. An alternative is to install sensing systems that can track objects without requiring a special marker. There exists a large variety of optical sensors for indoor positioning, as they are affordable and mimic the human visual system [18]. In addition to tracking, they also can be used to accomplish additional tasks, including gesture recognition or soft biometrics [5, 15]. However, the systems require an exposed lens, and there may be the limitation due to occlusion.

In this paper, we provide an introduction to electric potential sensors and ultrasound sensing, two sensing technologies that enable you to do the aforementioned tasks unobtrusively. Further, we introduce several prototypical applications that can be used in Smart Environments, to track the position or activities of several inhabitants. We present several use cases that have been developed in recent years and based on these we will discuss the opportunities and challenges of these marker-free technologies.

Electric potential sensors (EPS) are a modern variety of the first electrical measurement instrument, the versorium, which was developed around 1600 [9]. They measure the effect of charged bodies that move in the vicinity of a detector, caused by the Coulomb force. The human body carries a certain charge, the movement of which can be detected by EPS.

Ultrasonic systems use sound waves in a non-audible frequency range and evaluate how they are reflected within an environment. Despite their short range, it is possible to use ultrasonic technology for lot of indoor applications. Most commonly, they are used for range finders. However, we can also use the Doppler effect to detect activities within the detection range and use receiver arrays to build an indoor localization system.

Compared to camera-based systems, these methods are more privacy-preserving. These techniques only detect the position and certain activities of the users. For example, these sensors can not detect emotions, or which clothes a resident is wearing. They also cannot be used to spy on documents and objects. It was shown by [13] that consumers are more suspicious about using vision-based monitoring systems as compared to non-vision-based systems. According to [8] a feeling of security can enhance the quality of life. Therefore, it is clearly preferable to use non-optical systems inside a standard living environment.

2 Theoretical Background

In this section, we would like to present two used technologies that we use for our work. The theoretical foundations have been known for a long time. Now, we try to invent the appropriate use-cases by exploring the advantages of these existing sensing technologies.

2.1 *Electric Potential Sensing*

Electric field sensors use passive sensing methods to pick up changes in ambient electric potential generated by human movements. They can detect the movements of a human body and, under certain conditions, its presence at distances up to room scale [25]. They work through any non-conductive material and are especially suited for integration into everyday objects [31]. An array of electric potential sensors can be used to detect passive interaction patterns, e.g., the presence of a person, as well as active interaction, e.g., gestures performed by a user over an equipped surface [33].

EPS have been investigated in the domain of contact and remote ECG measurements [26, 27]. These approaches are realized with sensors measuring electric potential in a differential configuration. By comparing the electric potential measured by two or more sensors, it is possible to reconstruct the ECG signal, even in an unshielded environment.

A good source that gives a thorough overview of the physical principles of EPS can be found in Kurita or Ficker [6, 14]. We will briefly outline the basic principle. Generally, it is based on the potential change within the human body due to instantaneous change between the human body and its static environment. This could be caused by foot movements or hand movements. The induced potential redistribution within the human body can be clearly detected and measured with a remote electrode and sensor, e.g., placed on the ceiling or the door. When the foot leaves the ground, the electrical capacity between the human body and the ground changes. The human body can be considered as a virtual plate capacitor, which is represented by a serial connection of two different plate capacitors C_B and C_x, whereas C_B is the capacity between the human body and the static ground and C_x is the so-called step capacitor in [6].

$$\frac{1}{C} = \frac{1}{C_B} + \frac{1}{C_x}$$

The step capacity

$$C_x = \epsilon_a \cdot \frac{S}{x}$$

depends on the permittivity of the air ϵ_a, the distance x of the foot from the ground plane and the effective surface area S of the footwear on the ground. The body potential U_B evolved on the human body should be able to be calculated via the formula

$$U_B = \frac{Q_B}{C} = Q_B \cdot (\frac{1}{C_B} + \frac{x}{\epsilon_a \cdot S})$$

where Q_B is the charge on the human body and the definition of other mathematical symbols in the formula were already mentioned previously. This charge Q_B on the human body induced another charge Q on a distant measuring electrode. This induced charge Q will be measured and can be calculated using

$$Q = C_m \cdot (U_B - V)$$

whereas U_B is the body electrical potential caused by body movement and V stands for the known electrical potential of the measuring electrode. The distance of the placement of the measuring electrode from the human object is also of vital importance. If the measuring electrode is placed near the human object, the induced charge Q will be increased, since with decreasing distance from the measuring electrode, the measuring capacity C_m will increase and thus the induced charge Q as well. In the opposite, a departing measuring electrode will cause a decreasing induced charge Q. The induced electrical current I can be derived as in [14] described using

$$I = \frac{dQ}{dt} = C_m \cdot \frac{dU_B}{dt} = C_m \cdot Q_B \cdot (-\frac{x}{\epsilon_a \cdot S^2} \cdot \frac{dS}{dt} + \frac{1}{\epsilon_a \cdot S} \cdot \frac{dx}{dt})$$

This formula describes that the induced electrical current I depends strongly on the foot motion.

2.2 Ultrasonic Sensing

The most common application of ultrasound sensors in localization tasks are in pulse-based systems, where they are used measure the distance to a solid object, as described by Nishida et al. [24]. A short pulse will be sent towards the target, and the echo signal that travels back to the receiver will be detected. Using time-of-arrival-methods the distance of the reflecting object can be calculated. The disadvantage of this method is the lack of velocity information with regards to the sender; we discard the phase change information which is paired with the velocity data of the target object. Though modified systems use the time-distance-of-arrival to locate a single sender using multiple receivers, this method can be used in indoor localization [28]. A final group of methods uses the Doppler effect in ultrasonic frequency ranges for measurements. If a sender is moving towards a receiver, this will result in a shift towards higher received frequencies; if he moves away, the frequencies will be shifted to a lower range. This also applies to objects that reflect ultrasonic waves, such as persons who move in the detection range. We can further refine these to the waveform used for signal generation. There are continuous wave signals (CW) and frequency modulated periodic signal (FMCW) [10]. The first waveform allows you to easily detect the Doppler signal of captured echo signals. The second waveform modulates the excitation signal to gather more information about the reflected signals, such as the distance and the relative velocity.

The ultrasound frequency range begins at frequencies above 20 kHz, which is defined as the upper boundary of human perception. Functionality is used in the medical area, such as the ultrasound imaging of the inner organs of human beings [30]. Other industrial applications include liquid level measurement [11] or identification of the location of leakage in pipelines [34]. Newer methods use smartphones for ultrasound measurements, to sense activities or recognize gestures in the vicinity of the smart phone [7]. This opens the door to creating more interesting applications in the area of novel human-computer interaction systems. In our previous study, Fu et al. showed other possibilities of using ultrasonic measurement in lower operation ranges [7], as depicted in Fig. 4a, with a carrier frequency of only 20 kHz to detect some predefined activities performed in the vicinity of a smartphone. The advantage of using an unmodified smartphone to detect human activity has been proven to be its easy deployment on various smartphones without modification needed. Nandakumar et al. have shown, that it is possible to use ultrasound emitted by a smartphone to supervise persons for sleep apnea [22]. They implemented an FMCW method, to detect the chest movement of the test person. Gupta et al. used a laptop to detect gestures in front of the laptop by emitting a continuous signal, with a constant carrier frequency of 20 kHz [10].

3 Own Approaches

We have optimized and used the previously introduced technologies to develop localization and interaction solutions. We present some of them, in more detail, in the following section. These approaches are novel and offer many possibilities for further development.

3.1 Novel Use Cases for Electric Potential Sensing

Novel Use Cases for Electric Potential Sensing Our first EPS-based approach, for indoor localization, is a grid-based system. This method builds on the concept of Braun et al. for a capacitive sensing floor for indoor localization and fall detection [1]. Electrodes, consisting of thin insulated wires, are mounted with a grid size of 20 cm, below the floor covering but spread over the entire area. As depicted in Fig. 1, the two layers form a grid that covers the entire floor surface. This ensures that the foot, with an estimated length of 30 cm, will be on top of an electrode and can be detected reliably. The sensors are mounted on two sides, on the horizontal and vertical side of the grid. The changes of the electric field can be measured when an object approaches the sensor grid. Because each sensor has a global coordinate, it is possible to determine a 2D-location on the flooring. The system is very flexible, and several of these sensor fields can be combined into one global system. The length of the electrode wires can be adapted to the specific room layout. This means that even the most unusual floor plans can be covered with small subsystems. Because the sensors are placed only at the edges of the area, it is possible to mount these sensors

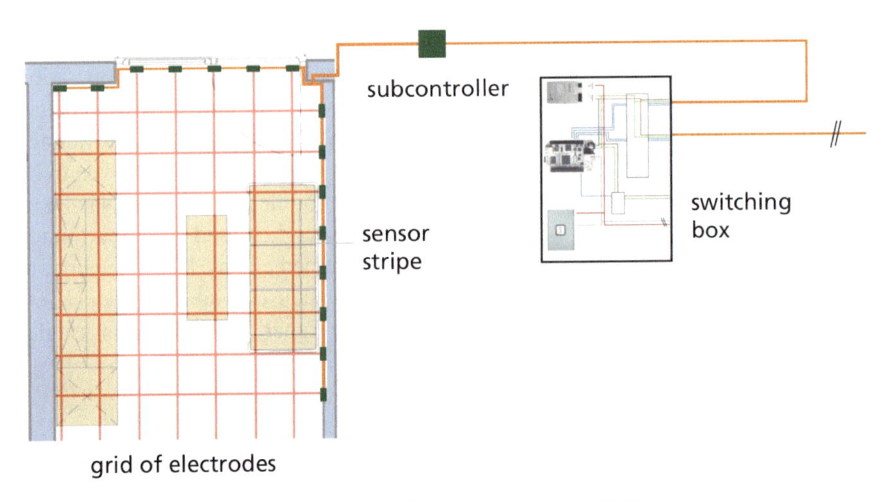

Fig. 1 Schematic structural drawing of the floor system based on EPS

behind baseboards or under covers of expansion joints. Thereby, easy maintenance without damaging the flooring is possible. The thin electrodes also allow simple and unobtrusive installation under almost any non-conductive floor covering. For example, it is possible to integrate the electrodes into the impact sound insulation of laminate or parquet. In contrast to other indoor localization systems, such as motion detectors, cameras, or marker-based systems, the user does not have to carry equipment at any time, and the system can be installed completely invisible. This approach also ensures that privacy of the tenants is respected. The system is designed to detect positions of one or more persons. This is possible due to the mesh structure and the intersection points. Through the use of clustering algorithms, multiple persons can be tracked. The algorithm is based on techniques that are frequently used in image processing [2, 19]. The main challenge is to detect new persons and to build new clusters. Unlike in image-processing regions can be defined, in which people appear or leave the system. Also, the specific characteristics of the human gait can help to differentiate between people. Steinhage et al. [32] have already shown, for a similar system, that these properties can contribute to distinguishing between humans and animals. In the presented system, it is also possible to detect falls. Therefore, we analyzed the limits of the area that is affected. The system is currently being installed as a prototype in a senior residence with over 20 apartments. One disadvantage of this scheme is that the electrodes have to be placed under the floor. Therefore, the system is not easy to install in existing houses or apartments (Fig. 2).

An ideal solution when retrofitting existing homes is to have a system that can be placed on ceilings. Dellangnol et al. examined this possibility and developed a prototype called platypus, which is a localization system that uses commercial electrical potential sensors, the PS25451 EPIC (electric potential integrated circuit) manufactured by Plessey Semiconductors, which is placed on the ceiling of the room. The system consists of groups of four electric potential sensors forming a rectangle [3].

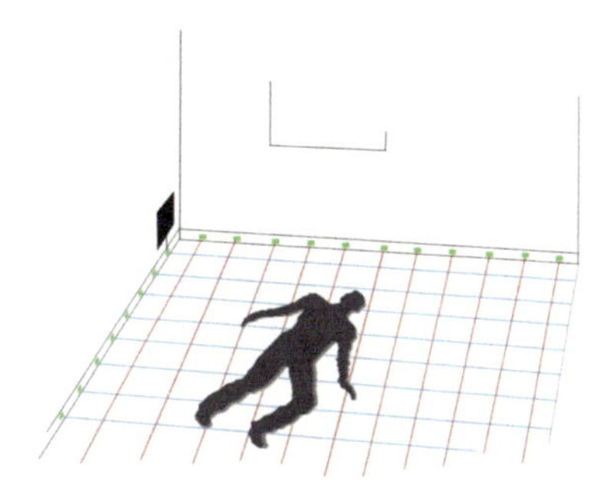

(a) A fall is detected by the number of electrodes and the type of signal

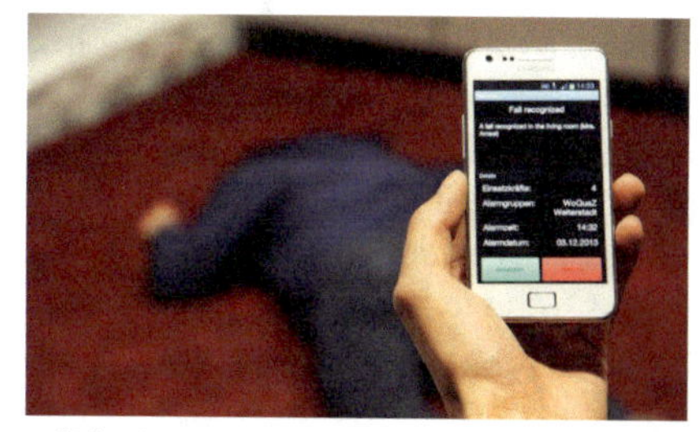

(b) Global coordinates make it possible to alert helpers and to find the fallen person

Fig. 2 Detection of emergencies by a smart floor

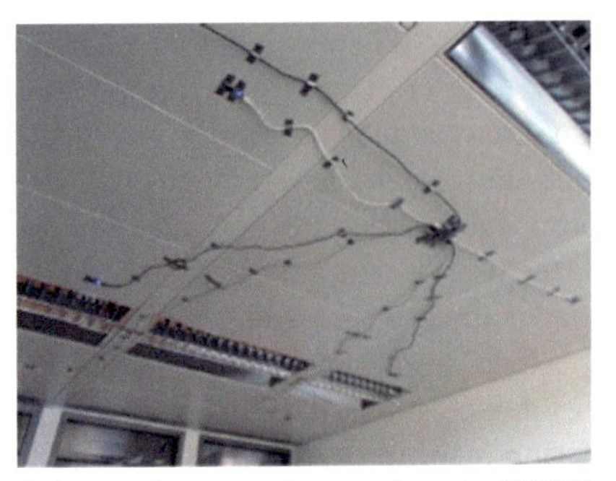

(a) The layout shows the six EPIC sensors placed on the ceiling to cover the test room [3]

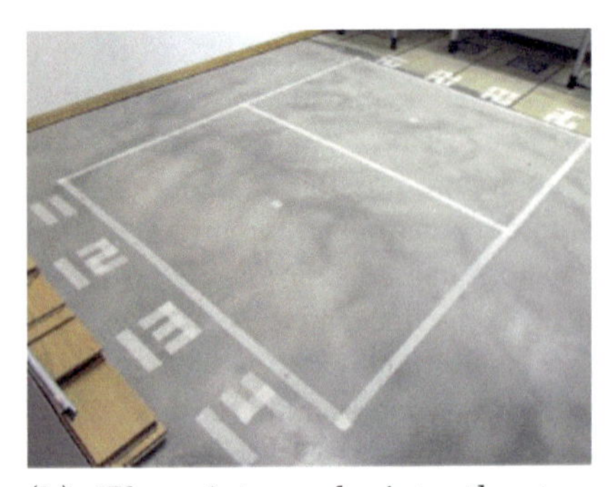

(b) The picture depicts the two sensor cells covering a total area of $5\,\text{m}^2$ [3]

Fig. 3 Those figures show the setup of the experiment conducted using our localization system with electric potential sensors

Installing several of these groups side by side allows for easily extending the system to cover large areas. The localization method is entirely passive and computationally inexpensive. Given that the calculations relies rely on the movement of the human body, a person can only be localized while walking. The system was tested in a controlled setting with 30 subjects (27 male, 3 female, min age 16, max age 36), who were asked to walk along straight paths under an array of six EPIC sensors (Fig. 3). The average localization error was found to be under 20 cm. Further studies are currently conducted to demonstrate localization and identification of persons in a more realistic setting.

We are also trying to explore other application areas of electric potential measurement technology. Sometimes the exact position of a person in a room is not needed. In many cases, it is enough to decide if and how many persons are in a room. For this use case, we placed the electric potential sensors hidden on the side of the door, to realize an unobtrusive person counter. We measure the variation of electric field induced by a person walking into or out of the doorway to update the person count. For this approach, we used two electric field sensors. Their electrodes were arranged in a v-shape. The electrodes were then placed near the door frame. This way, a person passing through the door would first activate one sensor, then both at the same time while in the middle of the doorway, then again only a single sensor when moving away. Owing to its low energy consumption and small size, such a system can easily be added to existing environments.

3.2 Novel Use Cases for Ultrasonic Applications

In this paper, we would like to introduce two more use-cases that we would like to implement using ultrasound technology. To start with, we aim to develop an

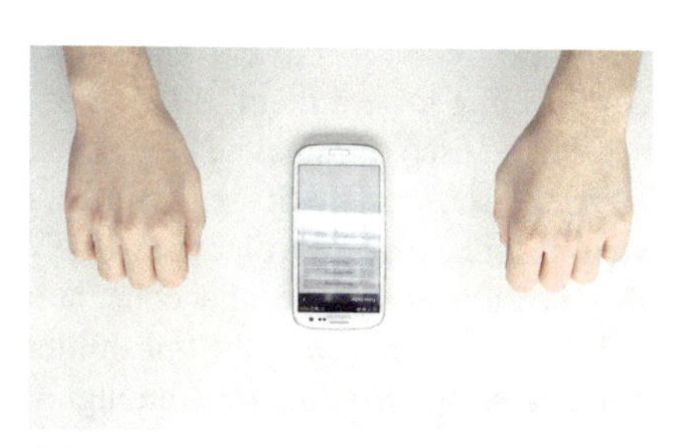

(a) Gestures Recognition using ultrasound measurement on smartphone. The depicted figure is taken from [7]

(b) Gestures Recognition using ultrasound measurement on laptop. The depicted figure is taken from [10]

Fig. 4 Those figures show the possible use-cases of ultrasonic technology in the application area of human-computer interaction

ultrasonic array system to perform indoor localization. Here, we profit from the knowledge of well-studied conventional radar technologies called the digital beamforming (DBF) technique with ultrasound transceiver array system, where we try to cover a large area simultaneously with high-resolution cells. It is a single-output and multiple-input system, in which one transmitter is continuously sending an output signal while the receiving arrays collect echo signals simultaneously to localize the subject located inside the sensing area. Similar indoor localization system using ultrasonic systems have been created by Mandlik et al. and Naghibzadeh et al. [17, 21]. Unlike our sensing technique, Naghibzadeh et al. realized indoor localization within a room by using ultrasound receiver arrays installed on the ceiling. They analyze the reflected signals to generate a baseline and use a time-distance method to identify the location of signals that deviate from the detected background (Fig. 4).

Another aspect to improve or simplify everyday life is a better motion detector using ultrasound waves. Ordinary motion detectors nowadays commonly use passive infrared sensors [12]. They are only able to tell that there is a movement in the vicinity of the sensor but are not able to gather information about the direction of the movement. Radar-based systems perform well in indoor environments due to the ability of radio waves to propagate through walls [29]. Gupta et al. have shown that it is possible to use WiFi signals to detect whole body activities even through walls. However, these systems are quite expensive due to the hardware components needed [35]. To overcome this problem, we propose an ultrasonic system that is inexpensive yet provides information about the direction of the movement. This additional feature, besides the conventional movement information, could strongly benefit the smart living applications and energy-saving aspect within the home automation domain. As an example, the light should only be switched on if the user intends to pass through an entrance and not just walk by. Furthermore, the light should be switched off as soon as the user leaves the room. Other possible use-cases can be thought of. While using ultrasonic sensing technology, this goal can be achieved by applying a time-difference-of-arrival (TDOA) method. By placing two receivers close to each other, we can measure the time difference of a signal traveling toward them. If the user is approaching the receivers, the signal difference can be used to calculate the approximate angle and direction of approach relative to the sensor. The first experimental setup using one ultrasound transmitter and two ultrasound receivers placed at a close distance of 2.5 cm beside it has shown promising results and proves this concept to be well suited for this kind of application. At a distance of 1 m toward the transmitter, the reflected signal energy from the movement of a person is still strong enough to be able to extract the directional information.

4 Outlook and Conclusion

In this paper, we have shown two sensing technologies that can ubiquitously sense human activity in an indoor area and are suitable for the next generation of indoor localization systems. Using EPS, we have shown a technology in a new context, which offers much potential for further applications. As an example, we present three systems that we have currently implemented: a floor-based system, a ceiling-mounted system, and a simple single tracker for proximity sensing. Ultrasonic sensing is a little more established, for which we have demonstrated two new use cases: beamforming for user tracking and a second approach for activity recognition. All these systems are well-suited for Smart Living environments as they are unobtrusive and discreet, and they preserve the privacy of the users.

As future opportunities, we would like to explore new applications for these technologies in smart environments. We will also try to identify further characteristic features that can be used to allow the identification of individuals. The EPS signal is, unfortunately, highly dependent on factors such as shoes or the static charge of the person. Therefore, we will primarily use behavioral characteristics as paths or walking patterns. Due to the limitations of each sensor on its whole, a possible solution to improve performance and to overcome these limitations is to use sensor fusion. Due to the unique advantage of low power consumption, EPS is also particularly suitable for large-scale deployment and the integration into the Internet of Things.

References

1. Braun, A., Heggen, H., Wichert, R.: Capfloor–a flexible capacitive indoor localization system. In: Evaluating AAL Systems Through Competitive Benchmarking. Indoor Localization and Tracking, pp. 26–35. Springer (2012)
2. Breitenstein, M.D., Reichlin, F., Leibe, B., Koller-Meier, E., Van Gool, L.: Online multiperson tracking-by-detection from a single, uncalibrated camera. IEEE Trans. Pattern Anal. Mach. Intell. **33**(9), 1820–1833 (2011)

3. Dellangnol, X.: Indoor localization based on electric potential sensing. Master's thesis, Darmstadt, TU, Master Thesis (2015), 81 p
4. Demiris, G., Rantz, M.J., Aud, M.A., Marek, K.D., Tyrer, H.W., Skubic, M., Hussam, A.A.: Older adults' attitudes towards and perceptions of smart home technologies: a pilot study. Inform. Health Soc. Care **29**(2), 87–94 (2004)
5. Demirkus, M., Garg, K., Guler, S.: Automated person categorization for video surveillance using soft biometrics. In: SPIE Defense, Security, and Sensing, pp. 76670P–76670P. International Society for Optics and Photonics (2010)
6. Ficker, T.: Electrification of human body by walking. J. Electrostat. **64**(1), 10–16 (2006)
7. Fu, B., Karolus, J., Grosse-Puppendahl, T., Hermann, J., Kuijper, A.: Opportunities for activity recognition using ultrasound doppler sensing on unmodified mobile phones. In: iWOAR 2015, p. 10. Association for Computing Machinery (ACM), ACM Press, New York (2015)
8. Gabriel, Z., Bowling, A.: Quality of life from the perspectives of older people. Ageing Soc. **24**(05), 675–691 (2004)
9. Gilbert, W., Mottelay, P., Wright, E.: William Gilbert of Colchester, Physician of London: On the Load Stone and Magnetic Bodies. Wiley (1893)
10. Gupta, S., Morris, D., Patel, S., Tan, D.: Soundwave: using the doppler effect to sense gestures. In: Proceedings of the SIGCHI Conference on Human Factors in Computing Systems, pp. 1911–1914. CHI '12 (2012)
11. Hao-hao, H., Jun-Qiao, X.: A method of liquid level measurement based on ultrasonic echo characteristics. In: 2010 International Conference on Computer Application and System Modeling (ICCASM), vol. 11, pp. V11-682–V11-684 (2010)
12. Keller, H.J.: Advanced passive infrared presence detectors as key elements in integrated security and building automation systems. In: Security Technology, 1993. In: 1993 International Carnahan Conference on Security Technology, Proceedings. Institute of Electrical and Electronics Engineers, pp. 75–77. IEEE (1993)
13. Kirchbuchner, F., Grosse-Puppendahl, T., Hastall, M.R., Distler, M., Kuijper, A.: Ambient intelligence from senior citizens perspectives: understanding privacy concerns, technology acceptance, and expectations. In: Ambient Intelligence, pp. 48–59. Springer (2015)
14. Kurita, K.: New approach to touch sensing technique based on measurement of current generated by electrostatic induction
15. Lee, H.K., Kim, J.H.: An hmm-based threshold model approach for gesture recognition. IEEE Trans. Pattern Anal. Mach. Intell. **21**(10), 961–973 (1999)
16. Lo, D., Mendonça, P.R., Hopper, A., et al.: Trip: a low-cost vision-based location system for ubiquitous computing. Pers. Ubiquitous Comput. **6**(3), 206–219 (2002)
17. Mandlik, M., Němec, Z., Vaňkát, T.: Real-time ultrasonic localization using an ultrasonic sensor array
18. Mautz, R., Tilch, S.: Survey of optical indoor positioning systems. In: 2011 International Conference on Indoor Positioning and Indoor Navigation (IPIN), pp. 1–7. IEEE (2011)
19. Milstein, A., Sánchez, J.N., Williamson, E.T.: Robust global localization using clustered particle filtering. In: AAAI/IAAI, pp. 581–586 (2002)
20. Mulloni, A., Wagner, D., Barakonyi, I., Schmalstieg, D.: Indoor positioning and navigation with camera phones. IEEE Pervasive Comput. **8**(2), 22–31 (2009)
21. Naghibzadeh, S., Pandharipande, A., Caicedo, D., Leus, G.: Indoor granular presence sensing with an ultrasonic circular array sensor. In: 2014 IEEE International Symposium on Intelligent Control (ISIC), pp. 1644–1649. IEEE (2014)
22. Nandakumar, R., Gollakota, S., Watson, N.: Contactless sleep apnea detection on smartphones. In: Proceedings of the 13th Annual International Conference on Mobile Systems, Applications, and Services, pp. 45–57. MobiSys '15 (2015)
23. Nirjon, S., Liu, J., DeJean, G., Priyantha, B., Jin, Y., Hart, T.: Coin-gps: indoor localization from direct gps receiving. In: Proceedings of the 12th Annual International Conference on Mobile Systems, Applications, and Services, pp. 301–314. ACM (2014)
24. Nishida, Y., Hori, T., Murakami, S., Mizoguchi, H.: Minimally privacy-violative system for locating human by ultrasonic radar embedded on ceiling. In: 2004 IEEE International Conference on Systems, Man and Cybernetics, vol. 2, pp. 1549–1554 (2004)

25. Prance, H., Watson, P., Prance, R.J., Beardsmore-Rust, S.T.: Position and movement sensing at metre standoff distances using ambient electric field. Measur. Sci. Technol. **23**(11), 115101 (2012)
26. Prance, R.J., Beardsmore-Rust, S.T., Watson, P., Harland, C.J., Prance, H.: Remote detection of human electrophysiological signals using electric potential sensors. Appl. Phys. Lett. **93**(3) (2008)
27. Prance, R., Debray, A., Clark, T., Prance, H., Nock, M., Harland, C., Clippingdale, A.: An ultra-low-noise electrical-potential probe for human-body scanning. Measur. Sci. Technol. **11**(3), 291–297 (2000)
28. Priyantha, N.B.: The cricket indoor location system. Ph.D. Thesis, Massachusetts Institute of Technology (2005)
29. Pu, Q., Gupta, S., Gollakota, S., Patel, S.: Whole-home gesture recognition using wireless signals. In: Proceedings of the 19th Annual International Conference on Mobile Computing and Networking, pp. 27–38. MobiCom '13, ACM, New York (2013). doi:10.1145/2500423.2500436
30. von Ramm, O.T., Smith, S.W.: Real-Time Volumetric Ultrasound Imaging System (1990)
31. Rekimoto, J., Wang, H.: Sensing gamepad: electrostatic potential sensing for enhancing entertainment oriented interactions. In: Extended Abstracts on Human Factors in Computing Systems, pp. 1457–1460 (2004)
32. Steinhage, A., Hoffmann, R., Lauterbach, C.: Automatische unterscheidung von personen und haustieren auf dem assistenzsystem sensfloor. In: AAL-Kongress 2015. VDE VERLAG GmbH (2015)
33. Steinhausen, N.: Applications of the electric potential sensor for healthcare and assistive technologies. Ph.D. Thesis, University of Sussex (2014)
34. Wang, X.J., Lambert, M.F., Simpson, A.R., Vitkovsky, J.P., et al.: Leak detection in pipelines and pipe networks: a review (2001)
35. Yarovoy, A., Ligthart, L., Matuzas, J., Levitas, B.: Uwb radar for human being detection. IEEE Aerosp. Electron. Syst. Mag. **21**(3), 10–14 (2006)

Technology Supported Geriatric Assessment

Sandra Hellmers, Sebastian Fudickar, Clemens Büse, Lena Dasenbrock, Andrea Heinks, Jürgen M. Bauer and Andreas Hein

Abstract Healthy aging is a core societal aim especially regarding the demographic change. But with aging, functional decline can occur and this is a major challenge for health care systems. For the evaluation of the health of older adults and the identification of early changes associated with functional and cognitive decline, clinical geriatric assessments are a well-established approach. Ideally, the assessments should take place at home of the older adults or even in their daily life, to get an unbiased functional status. Therefore, we introduce a technology supported geriatric assessment as an intermediate step to a home-assessment or in future to sensor-based-assessments in daily life. Beside various ambient sensors, a sensor belt is used during the assessments and for 1 week in the participants' daily life. We discuss the suitability of our measuring devices for an ambient home-assessment and evaluate the sensors in comparison to valid measurements. Thereby, we show that light barrier measurements achieve a high sensitivity and a good correlation to manual measurements through study nurses or physical therapists.

1 Introduction

Facing the challenge of demographic change, healthy aging is a core societal aim in our societies. With aging functional decline can occur and the older adults lose physical and mental abilities. A geriatric assessment is a well-established instrument to identify early changes associated with functional and cognitive decline, as they can occur in frailty or sarcopenia, which are common geriatric syndromes.

S. Hellmers (✉) · S. Fudickar · C. Büse · A. Hein
University of Oldenburg, Ammerländer Heerstraße 140, 26129 Oldenburg, Germany
e-mail: sandra.hellmers@uni-oldenburg.de

L. Dasenbrock · A. Heinks · J.M. Bauer
Department of Geriatrics, Klinikum Oldenburg, University of Oldenburg,
Rahel-Straus-Straße 10, 26133 Oldenburg, Germany
e-mail: lena.dasenbrock@uni-oldenburg.de

R. Wichert and B. Mand (eds.), *Ambient Assisted Living*,
Advanced Technologies and Societal Change,
DOI 10.1007/978-3-319-52322-4_6

Functional ability, physical health, cognition and mental health, and socioenvironmental circumstances of the older adults [1, 2] are evaluated in such an assessment.

It is important to recognize early changes of the functional status, because early interventions can improve or maintain physical function, reduce the rate of falls and help the older adults to maintain their independence [3]. But it's a challenge to identify older adults at risk of functional decline at an early stage and to initiate preventive measures when they are needed most. Due to the high effort and costs of a manual functional assessment, technology support can enable more extensive and frequent or even long-term measurements. Consequently, this allows a more detailed and current perspective of the health of patients and, thus, interventions can be initiated earlier.

Most geriatric assessments take place in hospital or medical offices. But optimally, the assessments should take place at home of the older adults or even in their daily life. In this case, an unbiased functional status of the patients can be assessed. A home-assessment could possibly be realized by new technologies. One related major question, to be investigated, is: Can specific sensors deliver similar results as a functional assessment and reduce efforts and costs?

Within our project for primary prevention for healthy aging, a clinical screening study is carried out to identify possible indicators for a functional decline. The study aims at the development of delta measures (difference between the results of two measures with a period of 6 months between them) to predict the need for long-term care via low-cost consumer sensors.

To make an intermediate step to a home-assessment or in future to sensor-based-assessment in daily life, the assessments are supported by technology. In addition to ambient sensors such as light barriers, a portable force plate and the aTUG (ambient Timed Up & Go) system, a wearable sensor belt is used. Each assessment item shall be mapped with the sensor data of the wearable device and analyzed afterward. After this assessment, the participants are asked to wear the sensor belt for 1 week. This allows rather continuous measurements of participants' behavior and items of activities of daily living and is possibly the next step to a daily life assessment.

The functional assessments are performed with 251 participants, aged 70 and above, at baseline (t_0), after 6 months (t_1) and 24 months (t_2) (see Fig. 1). We aimed to obtain a representative sample in this age group in terms of sex and socioeconomic

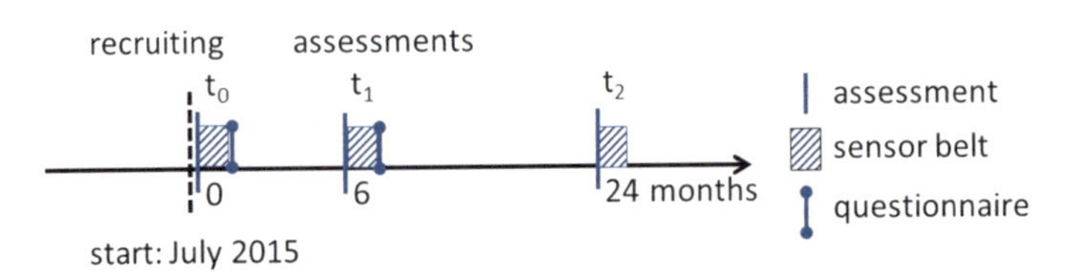

Fig. 1 Timeline of the clinical study. Participants pass the functional assessment 3 times within 2 years. After the assessment the participants wear a sensor belt for 1 week and record their activities in a diary. In addition a questionnaire will be completed by the older adults. At the measuring time t_0 the questionnaire contains questions about the socioenvironmental circumstances

status. At this time already 170 older healthy adults participated in our study. The mean age is 75.8 years, in a range from 70 to 87 years. 105 participants are female (61.8%) and 65 participants are male (38.2%).

Some important assessment items entailed in our study are not directly applicable in home-assessments. But the predictors, which are to be developed, for functional decline on the basis of technology measurements and the evaluation of ambient devices for integration into home environment, provide a foundation for these next steps.

2 Technology Supported Assessment

In order to analyze the suitability of ambient technologies for home assessment of functional status of the older adults, we use various ambient devices in our study. The combined assessment items of multiple common clinical assessments and the employed measurement devices for each item are described in this section.

Table 1 lists the considered measurement devices, the corresponding measured values and performed assessment items. Due to the high relevance of endurance, balance and strength for healthy aging, these assessment items are supported by ambient technology, to make the next step to future home-assessments. Light barriers are measuring the walking speed, leg strength and endurance during the Short Physical Performance Battery (SPPB), Frailty Criteria, Stair Climb Power Test (SCPT), Timed Up & Go (TUG) Test and 6 min Walk Test (6MWT). Mobility and strength are measured by the ambient Timed Up & Go (aTUG) system during the Timed

Table 1 Employed measurement devices within the assessment items

Category	Measurement device	Measured value	Assessment item
Ambient	Light barrier	Walking speed, endurance, leg strength	6MWT, SPPB, SCPT, TUG, frailty criteria
	aTUG system	Mobility, strength	TUG, SPPB
	Force platform	Balance, jump power/leg strength, mobility	CMJ, DEMMI, SPPB
Medical	Hand grip dynamometer	Grip strength	Frailty criteria
	Bio-impedance measuring device	Body composition, muscle mass	BIA
	Stadiometer	Height, weight, BMI	BMI
Body area	Sensor belt	Walking speed, balance, mobility, leg strength, jump power	6MWT, SPPB, SCPT, TUG, DEMMI, frailty criteria, CMJ

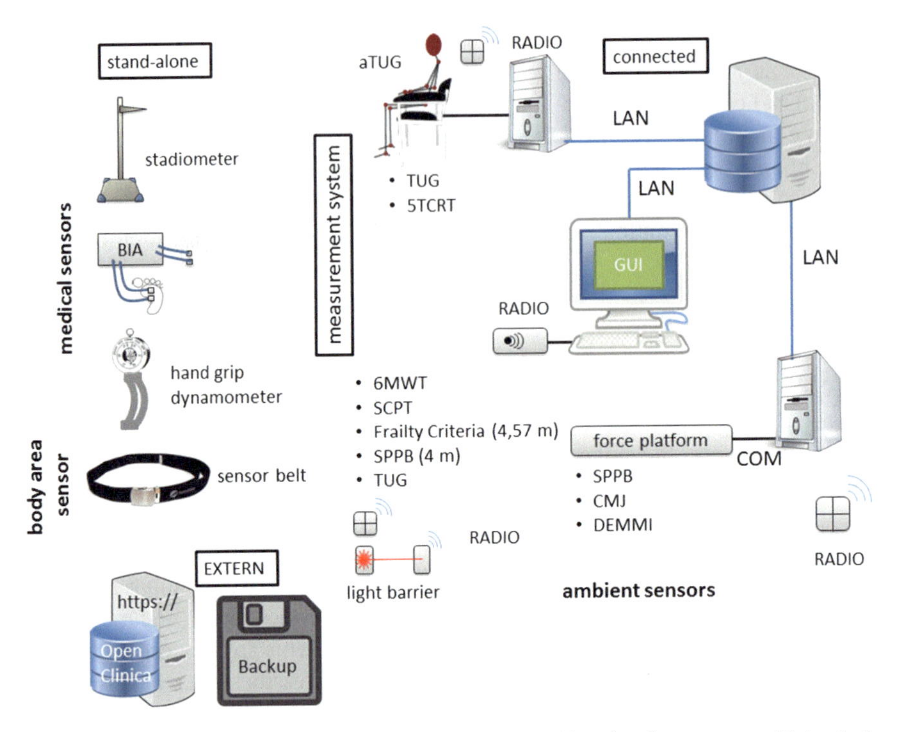

Fig. 2 Measurement system of the technologically supported functional assessment with its single components and the corresponding assessment items, data storage and documentation

Up & Go and the 5 Time Chair Rising Test (5TCRT), which is an item of SPPB. The force platform measures the balance and strength during the SPPB, the balance tests of the de Morton Mobility Index (DEMMI) and the Counter Movement Jump (CMJ). Special devices such as a bio-impedance measuring device, a stadiometer and a handgrip dynamometer are used for Bio-Impedance Analyses (BIA), measurements of the Body Mass Index (BMI) and for evaluating grip strength, which also provide important criteria for health. The wearable sensor belt monitors each assessment item. The data shall be analyzed and mapped to each assessment item.

The resulting complete measurement system with its specific components is shown in Fig. 2. Beside stand-alone devices like stadiometer, sensor belt, bio-impedance measuring device and hand-grip dynamometer the other devices are connected to a measuring computer. The measuring computer receives the data recorded by the aTUG system and the force platform via LAN and the radio signals of the light barriers via radio receiver. The distances for the walk tests in the measuring room are marked by tape on the floor. Various remote controls transmit start and stop signals for the different assessment items. The measurement program assembles the data, calculates the walk test duration on basis of the remote controls triggers and stores the data afterward. The documentation of the assessments is conducted using Open Clinica.

The ambient and wearable devices provide objective data and are independent of subjective influences like potential different reaction times of the experimenter. For a good reliability, it is necessary to verify the correct functionality and to ensure a high sensitivity. As a result, the early recognition of measurement errors is essential. A redundant system prevents the loss of data in such a case.

The measuring devices and the performed assessment items, listed in Table 1, are described in the following.

2.1 Ambient Sensors

Light Barriers

Light barriers are applied to measure walking time and walking speed for various distances at the specific assessment items. A walking test with a distance of 4 m is performed in the Short Physical Performance Battery (SPPB) [4], 4.57 m in the Frailty Criteria [5]. These tests are performed in the measuring room and are used to evaluate the walking speed.

The endurance is tested by the 6 min Walk Test (6MWT). During 6 min the participants are walking along the corridor over a distance of 20 m and return. One benefit of light barriers within the 6MWT is the fact, that the time of the turn at the end of the walking route and the associated deceleration can be excluded from the calculation of the walking speed.

The stair climb or leg power can be determined by the Stair Climb Power Test (SCPT) [6]. The stair climb power (SCP) can be calculated by following equation [7]:

$$SCP = m \cdot g \cdot h \cdot t^{-1}, \tag{1}$$

where m is participants mass in (kg), g is the gravity acceleration ($\approx 9.81\ \mathrm{ms}^{-2}$), and h is the height of the staircase in (m).

aTUG System

The ambient Timed Up & Go device (aTUG) [8, 9] includes light barriers, force sensors and a laser rangefinder. Thereby, it is able to detect the sit to stand cycle and to make detailed gait analyses. Figure 3 shows the aTUG system. Light barriers in a distance of 3 m from the aTUG system enable the analysis of the turn at the end of the walking route. The aTUG has been validated to measure reliably and precisely the total duration of TUG and durations of the single components of this sequence like—standing up, walking there, turning, walking back and sitting down—with mean error of only 0.05 s and mean standard deviation of 0.59 s using especially its force and range measurements [9]. The aTUG system supports the Timed up & Go test [10] and the 5 time chair rise test within the Short Physical Performance Battery (SPPB) [4].

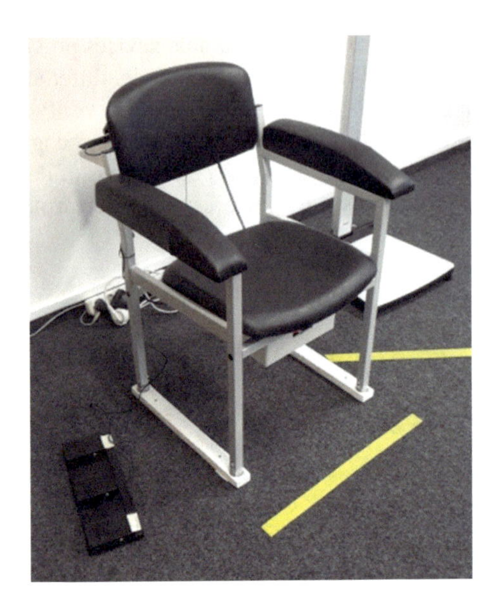

Fig. 3 The ambient Timed Up & Go (aTUG) device includes light barriers, force sensors and a laser range scanner. The device is able to detect the sit to stand cycle and to make detailed gait analyzes

Force Platform

To evaluate the balance, the mobility and the strength of the lower extremity the Short Physical Performance Battery (SPPB) [4] and the de Morton Mobility Index (DEMMI) [11] are performed by the participants on a portable force platform. The force platform is also used for the counter movement jump.

The AMTI AccuPower force platform is specified for jumping and power analyzes. Beside forces, power and velocity measurements the platform is also able to measure the center of pressure or mass and is suitable for balance analyzes. The AccuPower sensitivity is based on a 8900 N full scale F_z capacity and a 12 bit internal AD (±2048 bit) or about 4.3 N/bit.

2.2 *Medical Sensors*

Stadiometer, Bio-Impedance Measuring Device and Hand Grip Dynamometer

Before the technology supported assessment items start, the independent living skills of participants are collected by the Instrumental Activities of Daily Living Scale (iADL) questionnaire [12]. The Barthel index [13] is used to score the self-sufficiency and care dependency. In addition to the body mass index the body composition and muscle mass is determined by a bio-impedance analyzes (BIA). The body mass index is measured by a stadiometer (seca 285 measuring station) with an accuracy of ±2 mm in height measurements and of ±50 g in a weighting range between 25 and 100 kg and ±75 g in a weighting range between 100 and 150 kg. The

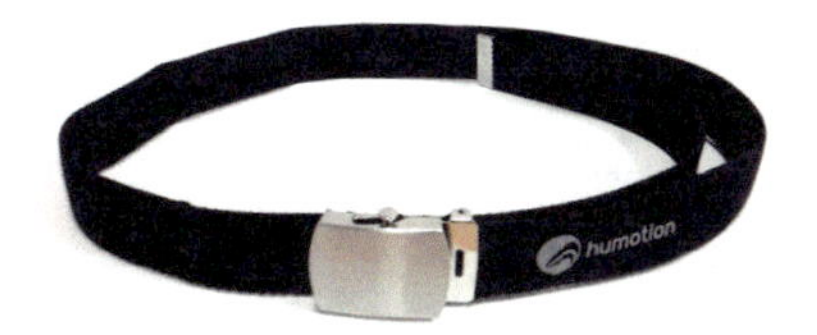

Fig. 4 Sensor belt with triaxial accelerometer, gyroscope magnetometer and barometer

bio-impedance measuring device has an accuracy of $\pm 2\,\Omega$ for the resistance R_z and $\pm 1\,\Omega$ for the reactance R_x.

The measurement of grip strength [14] with a hand grip dynamometer is an assessment item of the Frailty Criteria [5]. The hand grip dynamometer of JAMAR has an accuracy of ±5% of scale.

2.3 *Body Area Sensors*

Sensor Belt

During the functional assessment items the participants wear a sensor belt by Humotion (see Fig. 4). One aim of the study is to compare and to identify the single functional assessment items with the sensor data of the belt and to answer the question if the sensor belt provides equal or even more information or results as a conservatively measured geriatric assessment. Maybe wearing the sensor belt in daily life can be engaged instead of a functional assessment to get the same results in the future. The belt includes four sensor types. A triaxial accelerometer, which measures the acceleration force in $g \approx 9.81\,\mathrm{ms}^{-2}$ applied to the device on all three physical axes (x, y, and z), including the force of gravity. The gyroscope measures a device's rate of rotation in $deg \cdot s^{-1}$ around each of the three physical axes (x, y, and z) and the magnetometer measures the ambient geomagnetic field for all three physical axes (x, y, z) in μT. The barometer gauges the ambient air pressure in *hPa*. The sensors included in the belt are listed in Table 2. The belt is worn till the end of the assessment. Following the assessment, participants will be requested to wear the described sensor belt for 1 week as well as to keep an activity log to document activities like stair climbing or walking (Tables 3, 4 and 5).

Table 2 The specifications of the sensors included in the belt

Sensor	Frequency (Hz)	Resolution (bit)	Measuring range
Accelerometer	100	12	±16 g
Gyroscope	100	14	$\pm 2000\,\mathrm{deg} \cdot \mathrm{sec}^{-1}$
Magnetometer	100	12	±500 μT
Barometer	100	14	–

Table 3 Analysis of 4.57 m walk test

Model	B	Std. error	Beta	t	Sig.
(Constant)	0.135	0.171		0.789	0.432
Light barrier	0.911	0.045	0.904	20.164	0.000
Tester2	0.066	0.091	0.034	0.725	0.470
Tester3	0.075	0.084	0.042	0.891	0.375

Table 4 Analysis of 4.00 m walk test

Model	B	Std. error	Beta	t	Sig.
(Constant)	0.260	0.114		2.284	0.025
Light barrier	0.914	0.036	0.952	25.337	0.000
Tester2	0.099	0.048	0.077	2.047	0.043
Tester3	0.060	0.047	0.050	1.265	0.209

Table 5 Analysis of 6 min walk test

Model	B	Std. error	Beta	t	Sig.
(Constant)	70.177	22.09		3.177	0.002
Light barrier	0.841	0.053	0.852	15.914	0
Tester2	−1.365	9.483	−0.008	−0.144	0.886
Tester3	−8.397	8.822	−0.054	−0.952	0.344

On the basis of the measured data, a detailed analysis of the participants' behavior and their activities of daily living can be made. There are already existing algorithms detecting steps or activities like standing, sitting, lying or climbing stairs [15, 16].

3 Validation of the Light Barrier Measurement System

While most sensing devices are validated, the light barriers have to be validated due to the specifics associated with the placement. The results of the ambient measured assessment items and valid measurements by study nurses and physical therapists are compared in regard to correlation and differences. In our study light barriers are tested for indoor walking speed measurements and for evaluating the stair climb or leg power.

Method of Testing

The reliability and validity of the light barrier measurement system will be examined via correlations between already validated clinical tests and technology-based tests. Validity is measured by sensitivity and specificity. To determine the validity of our measuring system, we compare the technology-based test results with the results of conventional geriatric assessment items.

The measured values are continuous diagnostic variables. There are specific cut-off values for each assessment item, which divide the results into categories. To validate the light barrier system, we need to identify these categories to define the sensitivity and specificity for each item.

The cut-off value for the time over a distance of 4.57 m (Frailty Criteria) is 7 s for men with a height of ≤173 cm and 6 s for a height of >173 cm. The cut-off value for a women is 7 s with a height of ≤159 cm and 6 s with a height of 159 cm [5].

The cut-off value (lower limit of normal) for the 6 min walk distance d_{6min} for healthy adults is calculated by following reference equations from Enright et al. [17]:

$$\text{Men:} \quad d_{6min} = (7.57 \cdot h) - (5.02 \cdot a) - (1.76 \cdot w) - 309 - 153 \quad (2)$$

$$\text{Women:} \quad d_{6min} = (2.11 \cdot h) - (5.78 \cdot a) - (2.29 \cdot w) + 677 - 139, \quad (3)$$

where h is the height in *cm*, a is the age in years, w the weight in *kg*.

For example the minimum distance within the 6 min walk test of a 75 year old healthy man, weight 80 kg and height 180 cm should be 415.7 m to fit within the normal range.

Hinman et al. [18] mentioned that participants ascended and descended stairs at an average speed of 1.3 steps per second.

The analysis of the validity and the determination of the sensitivity and specificity is demonstrated with the example of the 4.57 m walk test. In order to assure the validity of both, the automatic and the manual measurement, results of 100 participants are compared regarding to their correlation and difference. Therefore, the first 100 error-free measurements are analyzed.

Results

According to the rule of thumb of the central limit theorem the chosen sample size of n = 100 indicates a normal distribution of the samples. This assumption is confirmed by the sample's normal distribution within the histograms and the q-q-plots, shown in Figs. 5 and 6, respectively. Therefore, all analyzed tests (4.57 m, 4.00 m and 6 min) are normally distributed.

The analysis shows that, due to the statistical significance values, there is a correlation between the stopwatch and light barrier measurements for the 4.57, 4.00 m SPPB and the 6 min walk test. Therefore, the results for the stopwatch measurement can be predicted on the basis of the light barrier measurements. The correlation can be described for the 4.57 m walk test by following regression line:

$$t_{stopwatch} = 0.911 \cdot t_{lightbarrier} + 0.135 \quad (4)$$

It has been further examined whether the different testers have an influence on the measured results. Three different experimenter conduct the functional assessments in

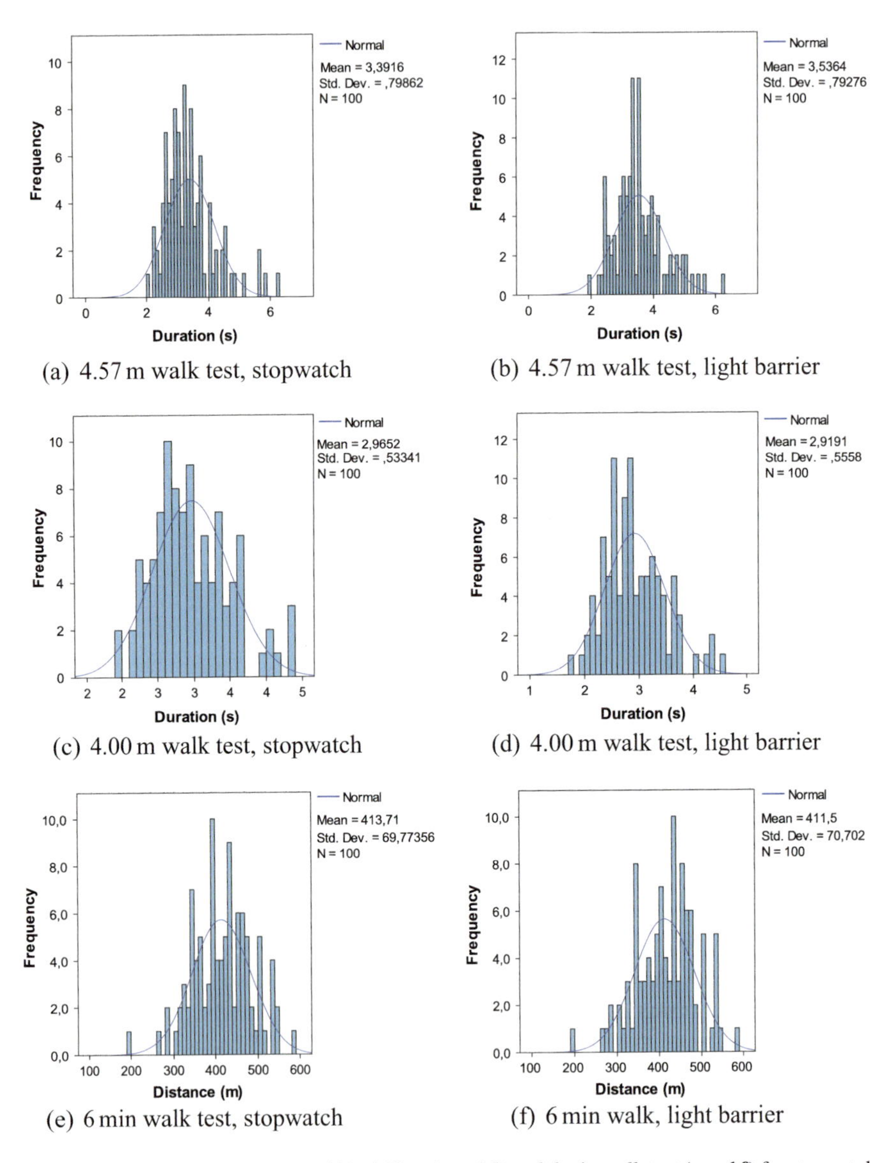

(a) 4.57 m walk test, stopwatch
(b) 4.57 m walk test, light barrier
(c) 4.00 m walk test, stopwatch
(d) 4.00 m walk test, light barrier
(e) 6 min walk test, stopwatch
(f) 6 min walk, light barrier

Fig. 5 Histograms of 4.57 m (**a** and **b**), 4.00 m (**c** and **d**) and 6 min walk test (**e** and **f**) for stopwatch and light barrier measurements. In (**a**–**d**) a time for the specific distance is measured in seconds (s). In (**e** and **f**) the measured value is the covered distance in meter (m) within 6 min. The mean value and standard deviation of each measurement is shown on the *right side* of each histogram

our study. There was no influence on the measurement for the 4.57 m and the 6 min walk test. But the analysis shows that tester 2 has a low influence for the 4.00 m walk test.

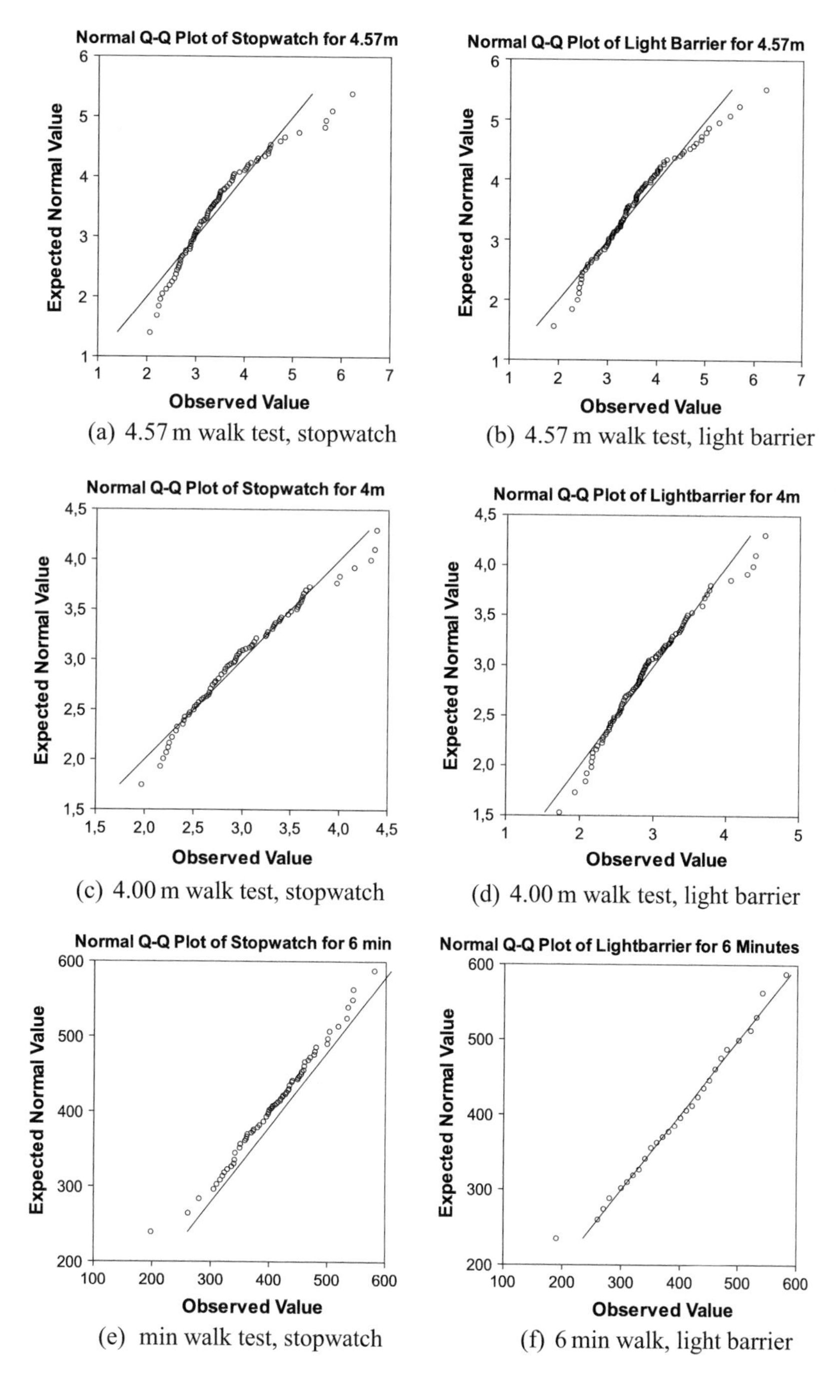

(a) 4.57 m walk test, stopwatch

(b) 4.57 m walk test, light barrier

(c) 4.00 m walk test, stopwatch

(d) 4.00 m walk test, light barrier

(e) min walk test, stopwatch

(f) 6 min walk, light barrier

Fig. 6 Normal q-q-plots of 4.57 m, 4.00 m and 6 min walk test for stopwatch and light barrier measurements

Table 6 Classification of the participants in frail and not-frail at the 4.57 m walk test on the basis of the cut-off values for this test

	Frail	Not-frail	Sum
Frail	1	0	1
Not-frail	0	99	99
Sum	1	99	100

Table 7 Validity of the light barrier measurement system

Test	Sensitivity (%)	Specificity (%)	κ	t-test	Correlation
Frailty (4.57 m)	100	100	1	0.2	0.9
SPPB (4 m)	100	100	1	0.55	0.93
6 min walk test	100	95.2	0.64	0.83	0.85

Table 6 shows the classification for 100 participants on the basis of the cut-off values [5] for the 4.57 m walk test. One senior of the 100 participants is classified as frail by the experimenter and the measuring system. None of the participants was falsely classified as frail by the measuring system. Therefore, the sensitivity is determined as 100% and the specificity as 100%. The kappa-coefficient κ is a measure of agreement between two sources, which is measured on a binary scale (i.e. category frail/not frail). A κ-coefficient between 0.80 and 1 characterizes a very good agreement and between 0.60 and 0.79 a good agreement. For the 4.57 m walk test the result for κ is 1. Table 7 shows the calculation for the assessment items supported by light barriers. Each light barrier supported assessment item performs with a good sensitivity. The specificity and the κ-coefficient at the 6 min walk test show no excellent conformity with respect to the gold standard. The reason for the false categorization of the participants can be attributed to the measuring setup. The light barriers are installed on the corridor with a distance of 10 m. Due to this fact, the accuracy of the measurement is about $\pm$10 m.

Optimization of Measurement System

To improve specificity and reliability of our system, consistency checks of the results are performed to identify incorrect measurements. If measured data passes a threshold value, because of too high or too low results, these measurements are automatically categorized as a false measurement. In addition, a redundant system records all light barrier signals during the whole assessment, so the results of each test can be reconstructed afterward, especially for missing start and stop signals by the remote control. In addition to acoustic signals at the start and end of the tests, an optical control system is implemented. Different colors show the status of the test in the graphical user interface of the measuring program and the experimenter can check if the test didn't start, didn't stop or which light barrier wasn't passed by the participants. To improve the specificity of the 6 min walk test, more light barriers will be installed to the second measuring time on the corridor. In this way, the accuracy will be enhanced up to ±5 m.

4 Applicability of the Devices for Ambient Home-Assessments

Considering the integrability of the applied ambient measuring devices at home, indoor walking speed could be measured with sufficient sensitivity by light barriers, which can be easily integrated at home [19, 20]. The aTUG system is an unobtrusive device for measuring sit-to-stand cycles in daily life or to analyze the walking speed by a laser rangefinder. Consequently, the aTUG system could replace normal chairs and measure these parameters during activities of daily living of the older adults. The force platform allows an evaluation of balance and strength. It is possible to integrate such a platform in the floor for example at frequently used places in home. Furthermore, force platforms or other large area sensor systems on the floor [21] are able to detect falls.

Isken et al. [22] discussed the practicability of clinical assessment tests in domestic environments and the use of ambient technology for home assessments. They described that tests have to be unobtrusive and should ideally be performed continuously without requiring patients to perform an explicit test. Some studies have already shown that assessment tests may be implemented without creating a test situation and thus can be implemented unobtrusively [23, 24]. Home-monitoring technologies were well-accepted by the participants in the GAL-NATARS study described in [25], where different ambient sensors like motion, power, vibration and door switch sensors are used for multimodal activity monitoring at home.

Beside the analysis at home, outdoor activities should be considered as well to achieve a high degree of significance to the functional status of older adults. Especially endurance measurements often take place outside. Therefore, devices with inertial units or sensors for vital signs are well-suited for identification of risks and for interventions [26]. In the clinical study described in this paper, the participants show a good acceptance of and compliance with the sensor-belt. None of the participants rejected to wear the belt. In the case of only 4 participants of 170 problems occurred. These problems were associated with the charging process of the battery of the sensor-belt. Next to highly accurate acceleration measuring devices like the sensor belt used in this study, other devices like smartwatches pedometers or GPS-loggers are possibilities for detecting indicators of functional decline in future. Such devices are affordable, so deployment in larger groups is not hindered by costs. They are easy to use in daily life and can also be used by technically less experienced people over a longer period of time [27].

5 Discussion and Further Work

This paper introduces a technology supported measurement system for a clinical functional assessment and evaluates its correctness in terms of correlation and difference to valid measurements by study nurses and physical therapists.

The system consists of a portable force platform, the aTUG system, a hand-grip dynamometer, a stadiometer and a bio-impedance measuring device as well as a light barrier measuring system. The light barrier system shows a high sensitivity and the results are in agreement with the valid stopwatch measurements, set as gold standard.

An advantage of light barriers and technology in general compared to manual measurements is the objective and equal measurement for each participant. If several study nurses respectively physical therapists perform an assessment, the potential different reaction times of each person influence the results. Another advantage lies in the fact that the time of the turn at the end of the walking route and the associated deceleration can be excluded from the calculation of the walking speed by using light barriers.

Furthermore, it is possible to get detailed information when using technology. The aTUG system detects the sit to stand cycle and makes detailed gait analyses besides measuring the duration of the TUG-test. The force platform delivers force, power and velocity values during the assessment items and allows a precision analysis of a jump action or balance tests. Thus, technology is able to support and enhance a clinical assessment. Moreover, we discussed the integrability and acceptance of different ambient measuring devices at home and their suitability for ambient home-assessments.

Beside ambient sensors, a wearable sensor belt is used in this study. In further work, the mobility data of the sensor belt will be analyzed. An aim of the study is to compare and to identify the single geriatric assessment items with the sensor data of the belt and to answer the question if the sensor belt provides equal or even more information or results as a conservatively measured assessment. Maybe, wearing this sensor belt for 1 week can be engaged instead of a geriatric assessment to get the same results in the future.

Acknowledgements This work has been funded by the German Federal Ministry of Education and Research (Project: Aequipa, grant no.: 01EL1422D).

References

1. Bassem, E., Higgins, K.: The geriatric assessment. Am. Fam. Phys. **83**(1), 48–56 (2011)
2. Clegg, A., Young, J., Iliffe, S., et al.: Frailty in elderly people. Lancet **381**(9868), 752–762 (2013)
3. Beswick, A., Rees, K., Dieppe, P., Ayis, S., et al.: Complex interventions to improve physical function and maintain independent living in elderly people: a systematic review and meta-analysis. Lancet **371**(9614), 725–735 (2008)

4. Guralnik, J., Simonsick, E., Ferrucci, L., et al.: A short physical performance battery assessing lower extremity function: association with self-reported disability and prediction of mortality and nursing home admission. J. Gerontol. **49**, M85–M94 (1994)
5. Fried, Linda P., et al.: Frailty in older adults evidence for a phenotype. J. Gerontol. Ser. A Biol. Sci. Med. Sci. **56**(3), M146–M157 (2001)
6. Bean, J., Kiely, D., LaRose, S., Alian, J., Frontera, W.: Is stair climb power a clinically relevant measure of leg power impairments in at-risk older adults? Arch. Phys. Med. Rehabil. **88**(5), 604–609 (2007)
7. Cataneo, D., Kobayasi, S., Carvalho, L., et al.: Accuracy of six minute walk test, stair test and spirometry using maximal oxygen uptake as gold standard. Acta Cirurgica Brasileira, Vpl. **25**(2), 194–200 (2010)
8. Frenken, T.: Technischer Ansatz zur unaufdringlichen Mobilitätsanalyse im Rahmen geriatrischer Assessments. VDI-Verlag (2013)
9. Frenken, T., Vester, B., Brell, M., Hein, A.: aTUG: fully-automated timed up and go assessment using ambient sensor technologies. In: 5th International Conference on Pervasive Computing Technologies for Healthcare, pp. 55–62 (2011)
10. Podsiadlo, D., Richardson, S.: The timed up and go. A test of basic functional mobility for frail elderly persons. J. Am. Geriatr. Soc. **39**, 142–148 (1991)
11. De Morton, N.A., Lane, K.: Validity and reliability of the de morton mobility index in the subacute hospital setting in a geriatric evaluation and management population. J. Rehabil. Med. **42**, 956–961 (2010)
12. Lawton, M., Brody, E.: Assessment of older people. Self-maintaining and instrumental activities of daily living. Gerontol. **9**, 179–189 (1969)
13. Mahoney, F., Barthel, D.: Functional evaluation. The barthel index. Md State Med. J. **14**, 61–65 (1965)
14. Mathiowetz, V., Kashman, N., Volland, G., Weber, K., Dowe, M., Rogers, S.: Grip and pinch strength: normative data for adults. Arch. Phys. Med. Rehabil. **66**(2), 69–74 (1985)
15. Kwapisz, J., Weiss, G., Moore, S.: Activity recognition using cell phone accelerometers. ACM SIGKDD Explor. **12**(2), 74–82 (2011)
16. Ravi, N., Dandekar, N., Mysore, P., Littman, M.: Activity recognition from accelerometer data. AAAI, pp. 1541–1546 (2005)
17. Enright, P., Duane, S.: Reference equations for the six-minute walk in healthy adults. Am. J. Respir. Crit. Care Med. **158**(5), 1384–1387 (1998)
18. Hinman, M., O'Connell, J., Dorr, M., et al.: Functional predictors of stair-climbing speed in older adults. J. Geriatr. Phys. Ther. **37**(1), 1–6 (2014)
19. Frenken, T., Steen, E., Brell, M., et al.: Motion Pattern Generation and Recognition for Mobility Assessments in Domestic Environments. AAL, pp. 3–12 (2011)
20. Steen, E., Frenken, T., Eichelberg, M., Frenken, M., Hein, A.: Modeling individual healthy behavior using home automation sensor data: results from a field trial. J. Ambient Intell. Smart Environ. **5**(5), 503–523 (2013)
21. Lauterbach, C., Steinhage, A., Techmer, A.: A large-area sensor system underneath the floor for ambient assisted living applications. In: Pervasive and Mobile Sensing and Computing for Healthcare, pp. 69–87. Springer, Berlin (2013)
22. Isken, M., Frenken, T., Frenken, M., Hein, A.: Towards Pervasive Mobility Assessments in Clinical and Domestic Environments. Smart Health, pp. 71–98. Springer (2015)
23. Frenken, T., Lipprandt, M., Brell, M., et al.: Novel approach to unsupervised mobility assessment tests: field trial for aTUG. In: Proc. 6th International Pervasive Computing Technologies for Healthcare (Pervasive Health) Conference (2012)
24. Stone, E.E., Skubic, M.: Passive, in-home gait measurement using an inexpensive depth camera: initial results. In: 2012 6th International Conference on IEEE Pervasive Computing Technologies for Healthcare (Pervasive Health), pp. 183–186 (2012)
25. Marschollek, M., Becker, M., Bauer, J.M., Bente, P., Dasenbrock, L., Elbers, K., et al.: Multimodal activity monitoring for home rehabilitation of geriatric fracture patients-feasibility and acceptance of sensor systems in the GAL-NATARS study. Inform. Health Soc. Care **39**(3–4), 262–271 (2014)

26. Meyer, J., Lee, Y.S., Siek, K., et al.: Wellness interventions and HCI: theory, practice, and technology. SIGHIT Rec. **2**(2), 51–53 (2012)
27. Meyer, J., Hein, A.: Live long and prosper: potentials of low-cost consumer devices for the prevention of cardiovascular diseases. J. Med. Internet Res. **2**(2) (2013)

Part III
Technology to Support Mobility

Gesture Controlled Hospital Beds for Home Care

S. Fudickar, J. Flessner, N. Volkening, E.-E. Steen, M. Isken and A. Hein

Abstract This article introduces a gesture-based user interface for hospital beds. This interface enables caregivers to focus on their patients and have both hands available for mobilizing and transferring them. Gestures are detected either via static (96% sensitivity) or dynamic gestures (67.5% sensitivity) and might be corrected by an extra repetition. Once gestures are correctly detected, caregivers can trigger bed movements via a foot switch as a hands-free operation, which as well functions as a dead man button. The evaluation of the usability through interviews with caregivers highlighted the system's general applicability, but as well some future challenges that have to be solved in order to achieve a system for every-day use.

1 Introduction

The increasing amount of people in need of daily stationary care is challenging for societies and results in an increasing demand and workload for caregivers [17]. Due to this increasing demand, a significant rise of the amount of caregivers can be expected (from 561,000 in 2010 to up to 900,000 in 2020) according to Pohl [17]. Similar trends apply as well for intensive care units which only in 2015 face a bedding increase of 8.5%.

While mobilizing or transferring patients, caregivers experience physical strain and face a high risk of severe injuries. Therefore, current hospital beds support caregivers by moving patients to an upright standing position. By lowering the required physical load for caregivers and increasing the safe handling of patients, these devices support even informal carers.

However, current devices require caregivers to continuously use a remote control during operation and thus bind one hand and limit caregivers' positioning and bind

S. Fudickar (✉) · J. Flessner · N. Volkening · E.-E. Steen · M. Isken · A. Hein
Division of Assistance Systems and Medical Engineering, School of Medicine and Health Sciences, University of Oldenburg, Ammerlaender Heerstr. 114-118, 26129 Oldenburg, Germany
e-mail: sebastian.fudickar@uni-oldenburg.de

R. Wichert and B. Mand (eds.), *Ambient Assisted Living*, Advanced Technologies and Societal Change,
DOI 10.1007/978-3-319-52322-4_7

their attention. Consequently, caregivers can only partially support patients while standing up or moving within the bed. Substituting the necessity to use remote controls with a rather hand-free interaction would enable full support of patients.

The use of gesture-recognition algorithms may represent a well suited alternative to remote controls as long as interaction is effortless and does not require user's full attention. This can be achieved by using gestures that are sufficiently intuitive to use and accurate to detect by the gesture recognition algorithms.

In order to investigate the practicability of a gesture-based control of hospital beds, we developed a gesture-based user interface for a hospital bed, validated its usability by analyzing the detection algorithms' accuracy and by conducting a qualitative user-study with caregivers.

2 Related Work

2.1 Enhanced Support Systems for Caregivers

The practicality of using gestures for user interaction has been shown to be supportive for caregivers in various projects. A lot of research is ongoing in the field of operating room support because touching devices is difficult in sterile environments. As an example Music et al. patented a system that can be controlled via gestures in a hospital setting to enable medical staff to access certain physiological characteristics of their patients without touching screens or buttons [14].

Gallo et al. used the Microsoft Kinect sensor to interact with medical images which is also suitable to be used in sterile settings [4]. Concepts of such systems can also be used to control different types of appliances.

The GeniAAL project [8] implemented a system, which enabled care-givers in care rooms to configure devices (such as lighting and bed-positions) for presets via gestures and confirmed the general practicality of this approach. The success of this approach is mainly associated with the following two gesture-related key challenges: On one hand, gestures and the mechanism for interaction must be acceptable to the user. On the other hand, the detection of the chosen gestures must be sufficiently accurate (in terms of sensitivity) by the algorithm. Consequently, both key-challenges will be discussed separately in the subsequent sections.

2.2 Usability Aspects

In order to investigate the acceptance and usability of the gesture controlled support system within GeniAAL, questionnaires have been conducted among caregivers and people in need of care [8]. The results indicate the suitability of gesture control support in the opinion of both user groups and clarified the advantage to use gestures that can be easily learned and memorized. Thereby, these questionnaires confirmed insights of the type of the suitability and generality of gestures, as being recognized by Kuehnel et al. [10]. Kuehnel et al. have shown that simpler gestures become better if they are memorable among different users and confirmed the general practicability of gestures for interaction with their surrounding. However, they have shown that the type and complexity of gestures is relevant for acceptance since only simple dynamic gestures are similar intuitive and memorable for most users. Among potential body parts, the hand is commonly accepted as most appropriate for gesture interaction [7].

Thus, guidelines for the development of gesture controls as the ones proposed by Preim und Dachselt [18] (as summarized in Table 1) emphasize adjustability, easiness and integration potential and are essential for establishing successful gestures for specific tasks and user-groups.

2.3 Gesture Recognition Algorithms

Various approaches have shown practicability for hand gesture recognition. Common approaches use typically either its contour [20] or the 3D representation in various levels of detail. Aside of the representation, gestures are either static (if only the position of fingers and palm is considered) or dynamic (if covering movements).

While various gesture recognition techniques (including inertial or contact sensor based ones) are available, the discussion herein will focus on visual (camera-based) gesture capturing techniques due to their suitability for the use-case.

The visual gesture-recognition consists of the following processing phases:

- Within the detection phase, image-information is analyzed and the hand (-position) is extracted. Among the visual techniques the use of depth-information is rather practical for the given scenario over the image-based data, since skin colors [11] might be inapplicable (e.g. if caregivers wear gloves to lower the risk of infection). For depth-based hand extraction the hand is identified either by its small proximity to the sensor [23], its contour [3], or through the modeling of the upper body skeleton [2].
- Within the tracking phase, the hand position determines hand and finger movements within the image. Typically either position-based 3D models [9] or movement-based [15] algorithms are used.
- In the feature extraction and classification phase, features are extracted based on the gathered information of the hand positions and movements. Based on the extracted features, gestures are then classified according to training vectors. Features are either skeleton-based or contour-based, depending on the approaches of the previous processing steps. The skeleton-based extraction can be achieved either via the angular characteristics among body parts [16] or by using the distance to a reference point, typically chosen to be the torso center [2]. The contour-based feature extraction is typically used for static gestures and can be characterized accurately by the centroid distance [23] or radial signatures [27]. The classification is related to the type of features available. While for static gestures

K-Nearest Neighbor (KNN) [6] or Scalable Vector Mashines (SVM) [22] have shown good results, for dynamic gestures Hidden Markov Models (HMM) or Dynamic Time Warping (DTW) algorithm [2, 13, 24] are applicable.

3 Usability Aspects

Due to the high significance of the effortless use of the intended interface, we identified relevant usability aspects via expert interviews and user interface design principles for the adaptability, simplicity and integration of interaction mechanisms (see Table 1) as summarized by Preim and Dachselt [18].

Regarding the herein intended gesture interface, selection of suitable gestures was essential to achieve an intuitive mapping of the desired functionality—which is represented by the hospital-beds' most desired degrees of freedom. These include vertical movements of the complete mattress, its head- and foot-sections as well as the complex movement for transferring lying patients into an upright position and vice versa (see Fig. 5). To identify adequate gestures for these functions, experts have been interviewed via both, paper-based questionnaires and qualitative interviews, regarding the following gestures preferences:

- Preference of static or dynamic gestures: Initial interviews indicated the preference of dynamic gestures.
- Preference for function-specific gestures (for both, static and dynamic gestures) as discussed in greater detail in Sect. 4.3.

4 Gesture Control

In order to achieve a gesture-based control of hospital beds, we intentionally considered only algorithms that use depth information as collected by sensors as found in the X-Box Kinect, because of being most applicable to the care scenarios (see Sect. 2). As discussed in the following subsections, we considered and tested two types of gesture recognition algorithms—namely static and dynamic gesture recognition, the latter reportedly having a better detection accuracy. In each case, we identified a generally applicable gesture-set by conducting focus groups (Sect. 4.3). In order to support the generation of user-specific gesture-sets and gesture-training, we as well implemented such functionality as described in Sect. 4.4.

Gesture classification systems typically consist of following processing steps (as shown in Fig. 1): Once gestures are assigned to bed functions, the system acquires frequently (with approximately 10 Hz for dynamic and 30 Hz for static gestures) depth images via a Kinect sensor.

In the subsequent processing-steps gestures are classified. However, since these processing steps are specific for static and dynamic gestures, they are discussed separately in the subsequent subsections.

Table 1 Considered design principles (based on [18]) and their applicability for the gesture-control

Design principle	Description of the principle	Implementation herein
Hand power and comfort	Avoidance of tiring postures/movements strenuous postures only for initiating gestures and relaxation as a gesture at the end	Quick gestures are used for selecting function and execution of function is triggered via foot switch
Fast, incremental and reversible actions	Quick performance of gestures. Easy correction of them	Foot switch enables incremental and fast use. Each gesture is reversible via its opposite gesture
Adequate feedback	Appropriate feedback from the system for performance of gestures	Visual feedback of current detected gesture/selected function
Ease of learning and memorization	Simplicity of gestures to assure rapid learning and memory	Gestures are chosen according to the intended bed movements
Consistency and symmetry	Similar gestures for similar functions	Gesture pairs: pointing up/down represent opposite movements; left/right arm represents opposite parts of bed
Good mapping of gestures on functions	Gestures representing functions. Define complex functions via experts or support customization	Gestures identified in interviews and representing bed movements
Adaptability to users and flexibility	Support customization of function/gesture pair mapping and individual gesture performance	Users can dynamically alter the mapping of gestures to functions
"Come as you are"	Minimize restrictions (such as body-worn tools) for performing of gestures	By using only ambient sensors, users can use the setup without preparation
Limiting amount of gestures	Implement only significant functions with gestures, to enhance memorability and technical detect-ability	Gestures limited to the bed's functionality and DoF

4.1 Static Gesture Recognition

The static gesture recognition algorithm has been implemented via the OpenCV library [1] and uses depth images of the Kinect V2 as input. The Kinect V2 higher depth image resolution (512×424 px) in comparison to the one of Kinect V1 (320×240 px) enables a more detailed analysis of hand shapes, which is assumed to enhance gesture recognition sensitivity.

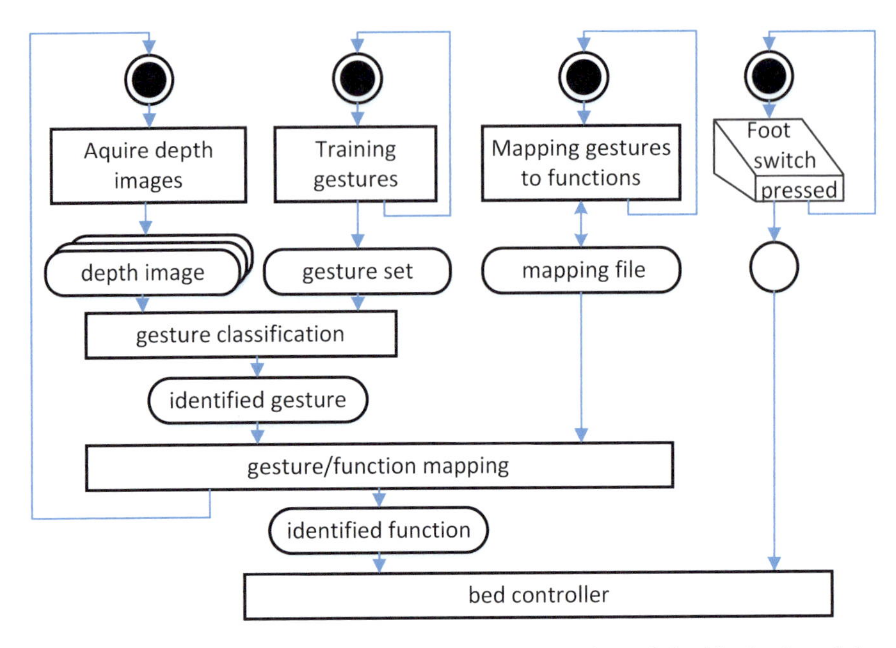

Fig. 1 General work-flow diagram of the gesture training, mapping and classification (encoded as fundamental modeling classes) and the triggering of bed functions

$$m_p q = \sum_x \sum_y x^p \cdot y^q \cdot f(x, y) \qquad p, q = 0, 1, 2, \ldots \tag{1}$$

The static gesture recognition algorithm consists of the following consecutive processing-steps:

1. Depth images are low-pass filtered with a maximal depth threshold of 2.2 m, in order to exclude potential noise and potential background objects.
2. Hand-shapes are calculated as the contour of remaining objects in the depth-image via the algorithm of Suzuki and Abe [26], which determines contours by following only the outermost borders as implemented by the *OpenCV findContours* functions. The algorithm results in point chains, that correspond to the contours.
3. Point chains of a length below a threshold of 5000 points are removed, to exclude irrelevant objects representing potential noise from further processing.
4. For each remaining object, the convex hull is determined via the algorithm of Sklansky [25].
5. Based on the calculated convex hulls, indentations are identified via the *convexityDefects* function of *OpenCV*. In order to identify only such indentations that might represent spread-out fingers, only such within a specific minimal/maximal length range are considered.

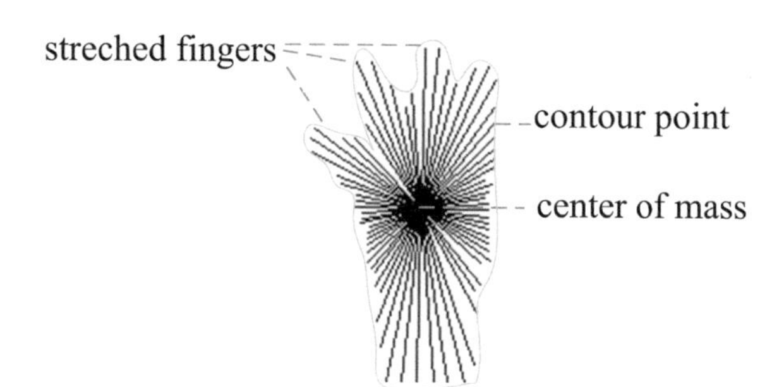

Fig. 2 Example for the static gesture recognition via distance values among the contour center of mass and specific contour points

6. Only objects that contain at least one relevant indentation are considered further on for the following feature extraction.
7. For each remaining object, the center of mass of the contour points is calculated via the *OpenCV moment* estimation function according to [28] (1). The feature of a gesture consists of its Euclidean distance to the object's center of mass. Thus, a distance vector is generated representing the distance in 5° steps. Figure 2 shows a representation for one gesture as the distances to the center of mass (represented by length of black lines).
8. Finally previous extracted features are evaluated via a Support Vector Machine (SVM). Therefore the feature vectors are labeled and added to a training-matrix, which is used for the SVM training. For SVM parametrization the *OpenCV train_auto* function has been used. Once an SVM is trained, gestures can be classified as well according to the newly trained feature.

4.2 Dynamic Gesture Recognition

The dynamic gesture recognition benefits from the skeleton modeling [2], which is integrated into the pre-processing of the Kinect sensor. Within the applied skeleton modeling, up to 20 3D-position skeleton joints of the human body (including the hand-joints) are localized within depth image streams [12]. Besides, the algorithm covers the detection of a simplified skeleton via an initial-pose (both arms above the head pointing upwards, so called psi-pose).

The classification of dynamic gestures is achieved via normalized distance variations from skeleton-junctions to a reference point over time [2, 21], representing the features. Since only arm and hand related gestures are considered, only upper-body parts (namely hands, elbows and shoulder) have been observed. As reference point the torso joint was chosen, since it has been shown to be robust regarding arm-movements [2, 21]. For features, variations over a timing window of 1.5 s are considered. These features are compared to features of trained gesture-sets via the Dynamic Time Warping (DTW) algorithm [13], which interprets time-based value sets and has been proven practical for dynamic gesture detection [2, 24]. The DTW

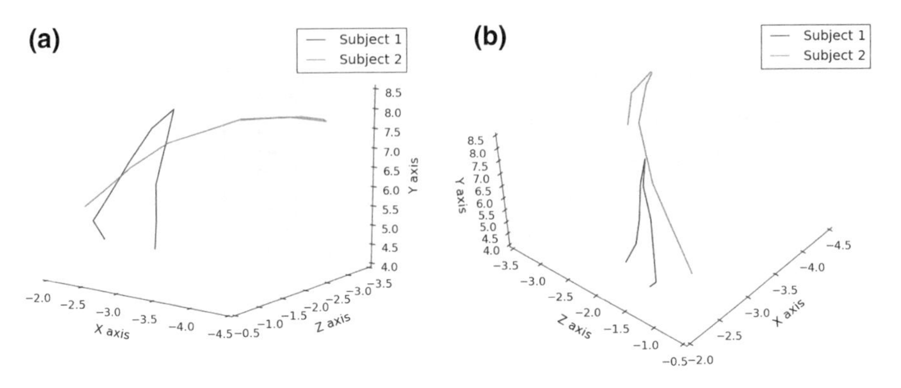

Fig. 3 Movement variances of two subjects executing the same gesture

algorithm was set up to classify gestures within movement-sequences as long as the distances surpass a threshold of 20. This threshold showed to support good sensitivity and specificity, by supporting typical variations of poses but excluding detection of unintended movements.

The described approach was chosen over other approaches (discussed in Sect. 2.3) such as angular distance due to its reported high sensitivity of up to approximately 97% [2] compared to respectively ca. 94% [16].

4.3 Gesture Sets

Since selection of suitable gesture-sets for static and dynamic gestures is essential for the system's usability, interviews with caregivers and unfamiliar users (representing informal carers) have been conducted. The results of these interviews indicated in accordance to the considered design-guidelines (as discussed in Sect. 2.2), that for easy memorability, gestures should represent associated functions' movements and that they should be performed with regional adjacency. Consequently, the static and dynamic gesture-sets that have been identified as appropriate in the interviews represent this fact.

While the side of the bed at which caregivers are approaching is relevant for the gesture-orientation, following description assumes that the bed is approached from the left (with left hand next to the head-segment). Due to the reconfigurability of gestures and gestures' symmetry, gesture-sets can be easily adjusted towards approaching from the right side.

For the identified relevant bed-movements, the **dynamic gestures** shown in Fig. 4 have been identified as most suitable, due to their regional proximity to the related segments. Most gestures (gestures 1–3) move along the vertical-axis and start at shoulder height: Moving segments upwards and rising the bed is represented by upward-leading gestures and respectively downward-leading gestures.

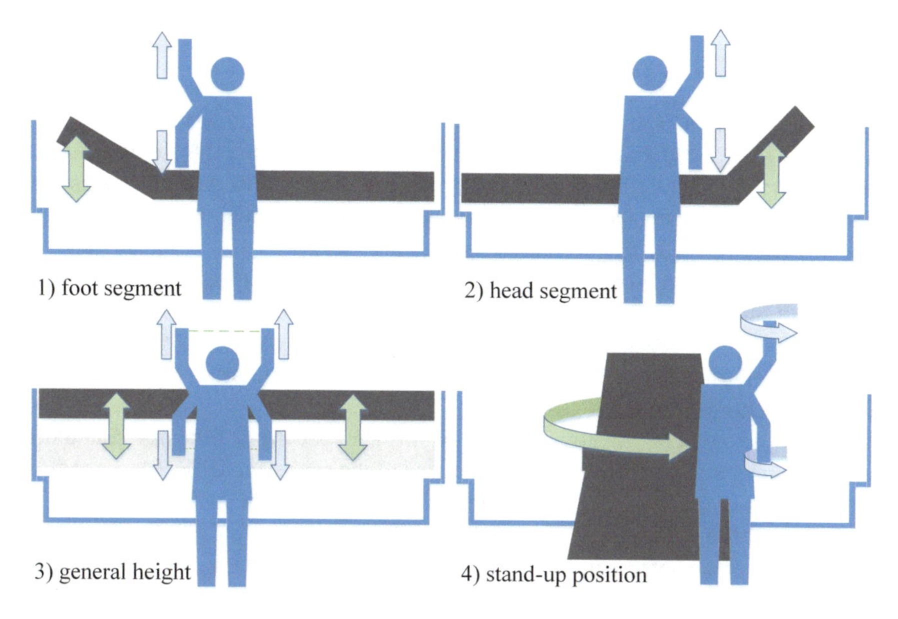

Fig. 4 Considered dynamic gestures and the related beds' degrees of freedom

Our gesture-set has similar gestures as the GeniAAL-project [8], for adjusting the bed's height, which as well identified the upward/downward hand-movement as most suitable in interviews with 33 experts.

The **static gesture**-set was chosen in accordance to the dynamic gesture-set, but by keeping the algorithmic limitations in mind (requiring fingers spread apart for recognition). For the static gesture set each degree of freedom is associated with a number-gesture. The static gesture set contains currently 45 training-features (as depth images) per gesture.

4.4 Training Gesture Sets

While classification of static gestures show suitable accuracy even without personalization, the system supports personalization for classification of dynamic-gestures. For dynamic gestures a *general dynamic gesture-set* was trained that consists of 1 sample for each considered gesture for 8 test-subjects. We assume that with this amount most common pose-variations will be covered. Thus, a sufficient classification sensitivity even for untrained persons should be achieved.

For classification of static gestures, a *general static gesture-set* currently containing 45 depth image samples per gesture, has been trained.

5 System

The overall system (shown in Fig. 5) integrates all necessary peripherals such as sensors (a Kinect and a foot switch), a display and a control unit (Arduino Uno and self-developed daughter-board) of the hospital bed, aside of the gesture recognition system. The components are subsequently described in accordance to their use in a typical work-flow.

The gesture recognition system, running on the server, records depth images via the Kinect, which is connected via USB. Currently detected gestures are visualized on the display, to give users instant feedback. In the current system version, the display is as well used as general interface for training and configuring the system (e.g. matching gestures to specific functions).

With a gesture being identified, users can trigger execution of the associated bed functions via a foot switch, which is connected to the server via USB. Since, control commands are send in accordance to the activation of the foot switch, the approach fulfills nearly the requirement of a dead man's switch.

As long as associated bed functions are activated via the foot switch, the server triggers the bed control unit to execute those. The bed control unit consists of a daughter-board on top of an Arduino Uno. The server is connect via an USB cable to the Arduino Uno, which generates a virtual serial port for the communication between both of them. The daughter board is controlled via I/O Ports of the Arduino Uno. Due to safety reasons optocoupler are used for the connection between the control unit and the hospital bed, which guarantee high quality galvanic isolation between both systems. Supporting additional bed-manufacturers would require only small adaptations of the daughter-board and eventually a limitation according to the degrees of freedom.

The hospital bed, in our case an INDREA B bed of the CareTec GmbH, Germany, has four degrees of freedom:

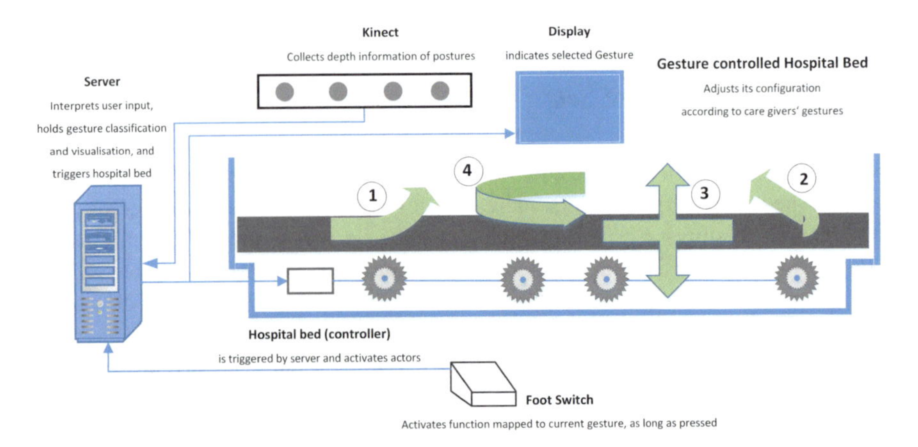

Fig. 5 Components and overall system of the gesture controlled hospital bed. Degrees of freedom (*green*): *1* foot-segment, *2* head-segment, *3* height of bed and *4* rotating to stand-up position

1. Angle and altitude of head and foot section
2. General altitude of the mattress
3. Sitting an lying position: a complex movement including rotating the mattress horizontally, raising the head-segment and lowering the foot segment, in order to bring the mattress into a rather vertical (standing) position.

The software stack uses the Robot Operating System (ROS) [19] as middle-ware and all software components (such as gesture recognition, UI and bed control unit) and peripherals (such as the foot switch and the bed control unit) are interconnected as ROS *nodes*.

6 Evaluation

The system has been evaluated regarding its gesture classification sensitivity and its usability, as described in the following subsections. In order to clarify the practicability of static and dynamic gestures, sensitivity of the classification has been evaluated separately for both gesture types.

6.1 Gesture Classification Sensitivity of Static Gestures

The classification accuracy of the static gesture recognition system (using the radial distance signature and a SVM) was evaluated with three subjects (one female, two males), for whom 15 recordings per gesture (allover 225 postures) have been recorded and considered. During the data acquisition, subjects performed gestures with their right hand in a distance of ca. 140 cm from the sensor—which represents a typical position of a caregiver standing approx. 20 cm in front of the bed.

Even though the static gesture classification was not optimized for specific users, the algorithm had a classification accuracy of 98.2% (221 out of 225 postures), as summarized in Table 2. So far, only the system's sensitivity has been evaluated—the specificity (in terms of the false-positive rate) has not yet been evaluated.

Table 2 Accuracy of static gestures with user-specific training. Intended gesture: horizontal and detected gesture: vertical

	1	2	3	4	5
1	**44 (98%)**	1	0	0	0
2	1	**44 (98%)**	2	0	0
3	0	0	**43 (96%)**	0	0
4	0	0	0	**45 (100%)**	0
5	0	0	0	0	**45 (100%)**

Table 3 Accuracy of dynamic gestures with user-specific training. Intended gesture: horizontal and detected gesture: vertical

	Upper right	Lower right	Upper left	Lower left	Upper circle	Lower circle
Upper right	62 (77.5%)	0	0	0	6	0
Lower right	9	65 (81.25%)	3	3	11	9
Upper left	0	0	51 (63.75%)	2	0	0
Lower left	5	5	2	67 (83.75%)	1	5
Upper circle	2	0	0	0	38 (47.5%)	0
Lower circle	0	0	0	0	2	39 (48.75%)
False positive summed	16 (20%)	5 (6.25%)	5 (6.25%)	5 (6.25%)	20 (25%)	14 (17.5%)
Not detected	2 (2.5%)	10 (12.5%)	24 (30%)	8 (10%)	22 (27.5%)	27 (33.75%)

6.2 *Gesture Classification Sensitivity of Dynamic Gestures*

In order to analyze the gesture classification sensitivity of the DTW algorithm, 10 recordings per gesture have been recorded for each of the eight test subjects. Consequently, for each of the 6 considered gestures (upper and lower right, upper and lower left, and upper and lower circle) 80 recordings have been used for evaluation. Each recording covered the gestured being performed twice. Classification sensitivity of dynamic gesture recognition with user-specific training of the classification algorithms has been trained via training sets for all eight subjects.

Combined 480 postures of eight test subjects (10 postures per gesture and subjects) have been evaluated. The cohort consisted of 6 males and 2 females. The average age was 37 years and ranged from 25 to 55 years.

Out of the 480 postures, 322 have been classified correctly (67.5%) as shown in Table 3. In addition, 93 postures (19.3%) could not been classified at all and 65 postures (13.5%) have been classified falsely, and thus represent false-positives. The false-positive recognition occurred mainly for the right upward gesture and both circle gestures where up to 25% have been miscategorized.

6.3 *Usability-Expert Interviews*

In order to evaluate if the developed system is supportive for caregivers, we conducted qualitative interviews with four caregivers. Within these we assessed their view on the systems usability (in terms of helpfulness, work simplification and intuitiveness of interaction) and critical points of optimization.

The interviews have been structured as follows: In the initial prototype demonstration, basic functionality of the hospital bed (including the uncommon stand-up functionality), the general aim of the gesture-based support-system and the relevant components (e.g. the use of the foot switch) were introduced to the interviewees. Consecutively, the gesture-control was executed and each gesture and function was once shown by the interviewer, who assured that the participants recognized the alteration of functionality via the feedback-mechanism. Afterwards, participants were given time to test the bed-control interface on themselves.

The sequel guideline-based interview via predefined questionnaires gave the following insights. All experts recognized the benefits of the gesture-based interface for hospital beds, compared to the remote-controls due to the following reasons:

- Having both hands free for interaction with the patient, since being able to trigger the beds' movements/functions via the foot switch.
- Lowering the risk of smear infection, due to overcoming the necessity of touching a remote-control.

The visual feedback mechanism indicating the gesture selection via a color-encoded status display was mainly perceived as suitable.

However, not all experts agreed that a gesture control may facilitate the everyday life of the nursing staff due to the following reasons: The complexity of some gestures (especially the circle gestures) was seen as problematic. Some experts have been skeptically if the complexity of the gestures might have a negative impact on the integration of the system in the nursing staff's work process. Also the appropriate placement of the visual sensors was seen challenging e.g. for nursing rooms in retirement homes, due to limited spaces and the required wiring.

The use of a foot switch was generally welcome as a control mechanism and was seen as suitable to prevent erroneous operations (e.g. if a wrong gesture was recognized). However, it was pointed out that the foot switch represents a potential source of danger (since stumbling over cables or the switch itself), which should be considered further.

7 Discussion

The developed gesture-recognition user interface for hospital beds overcomes the necessity of caregivers to continuously use a remote control while moving bed segments. By triggering the movements via a foot switch caregivers can use both hands and have full attention to support patients. In order to clarify the usability of the proposed user interface, we evaluated the system's recognition sensitivity and its usability.

The gesture recognition sensitivity for static gestures was sufficiently accurate (96% recognition rate) to assure an effortless use. For dynamic gestures the false positive rate reached up to 25% being significantly higher than the one of static gestures.

This represents a challenge, since users prefer dynamic gestures, while static ones are detected more precisely.

However, since false-positive classified dynamic gestures are neither associated to a single user or gesture, a repeated execution of the gesture results probably in a correctly detected gesture. The low sensitivity rates of the rather complex circle gestures (of 48 and 49%) might derive from the variations by subjects performing the gestures e.g. rotating in the wrong direction, or with different radius (as shown examplistically in Fig. 3). Consequently, the training-sets for these gestures should be further extended or these gestures should be even reconsidered in general, since their complexity was seen as problematic to memorize by the interviewees. Thus, the system supports exchanging specific gestures by the users themselves.

While integration of the user interfaces (namely the foot switch and the color-encoded status display) into the care workflow was seen as generally beneficial and welcomed (especially since freeing both hands and lowering the risk of smear infection, which has been shown to be a critical way of infection within hospitals and nursing homes [5, 29]), the following risks have been pointed out, that have to be taken into account. Herein, the wiring of the foot switch has been identified as a potentials risk (for patients falling). Furthermore, the positioning of the depth-sensor is challenging due to the space constraints and the required wiring.

Even though, the current system holds some enhancements of the care process, we have identified potential further optimizations that will increase the acceptance and applicability of the system in everyday care scenarios. The system has to be tested with additional users to increase the validity of this study.

References

1. Bradski, G.: OpenCV. Dr. Dobb's J. Softw. Tools (2000)
2. Celebi, S., Aydin, A.S., Temiz, T.T., Arici, T.: Gesture recognition using skeleton data with weighted dynamic time warping. In: Battiato, S., Braz, J. (eds.) VISAPP 2013—Proceedings of the International Conference on Computer Vision Theory and Applications, vol. 1, Barcelona, Spain, 21–24 February 2013, pp. 620–625 (2013)
3. Dhawan, A., Honrao, V.: Implementation of hand detection based techniques for human computer interaction. Int. J. Comput. Appl. **72**(17) (2013)
4. Gallo, L., Placitelli, A., Ciampi, M.: Controller-free exploration of medical image data: experiencing the kinect. In: 2011 24th International Symposium on Computer-Based Medical Systems (CBMS), pp. 1–6 (2011). doi:10.1109/CBMS.2011.5999138
5. Gerdes, S., Redlich, C., Yilmaz, M.: 4. Gesundheitsbericht 2015: Norovirus kompakt (2015). http://www.hannover.de/content/download/542635/12405677/file/Gesundheitsbericht_2015_Norovirus.pdf. Accessed 13 Nov 2015
6. Hasan, H., Abdul-Kareem, S.: Static hand gesture recognition using neural networks. Artif. Intell. Rev. **41**(2), 147–181 (2014). doi:10.1007/s10462-011-9303-1
7. Karam, M.: A Framework for Gesture-Based Human Computer Interactions. VDM Verlag, Saarbrücken (2009)
8. Keiser, T., Höß, O., Klein, B., Neuhüttler, J., Schneider, H., Vetter, T.: Gestensteuerung im Pflegeumfeld—Das Projekt GeniAAL: Grundlagen, Anwendungsfelder, Technologien und Erfahrungen. Books on Demand (2015). https://books.google.de/books?id=qerVBgAAQBAJ

9. Keskin, C., Kra, F., Kara, Y., Akarun, L.: Real time hand pose estimation using depth sensors. In: Fossati, A., Gall, J., Grabner, H., Ren, X., Konolige, K. (eds.) Consumer depth cameras for computer vision, advances in computer vision and pattern recognition, pp. 119–137. Springer, London (2013). doi:10.1007/978-1-4471-4640-7_7
10. Kühnel, C., Westermann, T., Hemmert, F., Kratz, S., Möller, S.: I'm home: defining and evaluating a gesture set for smart-home control. Int. J. Hum. Comput. Stud. **69**(11), 693–704 (2011). doi:10.1016/j.ijhcs.2011.04.005
11. Liwicki, S., Everingham, M.: Automatic recognition of finger spelled words in british sign language. In: Proceedings of the 2nd IEEE Workshop on CVPR for Human Communicative Behavior Analysis (CVPR4HB'09). In conjunction with CVPR2009, pp. 50–57. IEEE Computer Society, Los Alamitos, CA, USA (2009)
12. Microsoft Developer Network: Skeletal tracking (2015). https://msdn.microsoft.com/en-us/library/hh973074.aspx. Accessed 13 Nov 2015
13. Müller, M.: Information Retrieval for Music and Motion. Springer, New York (2007). doi:10.1007/978-3-540-74048-3
14. Music, D., Eghbal, D., Vargas, S.: User interface and identification in a medical device system and method (2010). https://www.google.com/patents/US7706896. US Patent 7,706,896
15. Park, S., Yu, S., Kim, J., Kim, S., Lee, S.: 3d hand tracking using kalman filter in depth space. EURASIP J. Adv. Signal Proces. **2012**(1), 36 (2012). doi:10.1186/1687-6180-2012-36
16. Pham, C.H., Le, Q.K., Le, T.H.: Human action recognition using dynamic time warping and voting algorithm. VNU J. Sci. Comput. Sci. Commun. Eng. **30**(3), 22–30 (2014)
17. Pohl, C.: Der zukünftige Bedarf an Pflegearbeitskräften in Deutschland: Modellrechnungen für die Bundesländer bis zum Jahr 2020. Comparative Population Studies—Zeitschrift für Bevölkerungswissenschaft **35**(2), 357–378 (2010)
18. Preim, B., Dachselt, R.: Interaktive Systeme: Band 2: User Interface Engineering, 3D-Interaktion, Natural User Interfaces, vol. 2. Springer Vieweg (2015). doi:10.1007/978-3-642-45247-5
19. Quigley, M., Conley, K., Gerkey, B., Faust, J., Foote, T.B., Leibs, J., Wheeler, R., Ng, A.Y.: ROS: an open-source robot operating system. In: ICRA Workshop on Open Source Software (2009)
20. Rautaray, S., Agrawal, A.: Vision based hand gesture recognition for human computer interaction: a survey. Artif. Intell. Rev. **43**(1), 1–54 (2015). doi:10.1007/s10462-012-9356-9
21. Rehrl, T., Blume, J., Bannat, A., Rigoll, G., Wallhoff, F.: On-line learning of dynamic gestures for human-robot interaction. In: 35th German Conference on Artificial Intelligence, KI 2012, Saarbrücken, Germany (2012)
22. Ren, Y., Zhang, F.: Hand gesture recognition based on meb-svm. In: International Conference on Embedded Software and Systems, 2009. ICESS '09, pp. 344–349 (2009). doi:10.1109/ICESS.2009.21
23. Ren, Z., Yuan, J., Zhang, Z.: Robust hand gesture recognition based on finger-earth mover's distance with a commodity depth camera. In: Proceedings of the 19th ACM International Conference on Multimedia, MM '11, pp. 1093–1096. ACM, New York (2011). doi:10.1145/2072298.2071946
24. Schramm, R., Jung, R.C., Miranda, E.R.: Dynamic time warping for music conducting gestures evaluation. IEEE Trans. Multimedia **17**(2), 243–255 (2015). doi:10.1109/TMM.2014.2377553
25. Sklansky, J.: Finding the convex hull of a simple polygon. Pattern Recognition Letters **1**(2), 79–83 (1982). doi:10.1016/0167-8655(82)90016-2
26. Suzuki, S., Abe, K.: Topological structural analysis of digitized binary images by border following. CVGIP—Graph. Mod. Image Proces. **30**(1), 32–46 (1985)
27. Trigueiros, P., Ribeiro, F., Reis, L.: Hand gesture recognition for human computer interaction: a comparative study of different image features. In: Filipe, J., Fred, A. (eds.) Agents and Artificial Intelligence, Communications in Computer and Information Science, vol. 449, pp. 162–178. Springer, Berlin (2014). doi:10.1007/978-3-662-44440-5_10

28. Wahl, F.M.: Digitale Bildsignalverarbeitung: Grundlagen, Verfahren. Springer, Beispiele (1984)
29. Winther, B., McCue, K., Ashe, K., Rubino, J., Hendley, O.: Contamination of environmental surfaces during normal daily activities of hotel guests with rhinovirus colds. In: 46th Annual ICAAC—Interscience Conference on Antimicrobial Agents and Chemotherapy, September 27–30, 2006, San Francisco (2006)

Assisted Motion Control in Therapy Environments Using Smart Sensor Technology: Challenges and Opportunities

Julia Richter, Christian Wiede, André Apitzsch, Nico Nitzsche, Christiane Lösch, Martin Weigert, Thomas Kronfeld, Stefan Weisleder and Gangolf Hirtz

Abstract In the following decades, the European population will be steadily growing older, which causes serious problems, especially with regard to the health sector. Problems are further aggravated by the lack of personnel resources. Even now, the number of therapists is not sufficient to supervise the increasing number of patients during their rehabilitation process. At this point, technical systems can support both patients and therapists in order to ensure the quality of rehabilitation. In this study, we review recent developments in the field of feedback-based therapy systems and identify needs that have not been satisfied thus far. On the basis of these findings, we introduce a technical system for assisted motion control in real therapy applications and discuss possible solutions in order to encounter current deficits. We thereby address aspects, such as sensor technologies, approaches for capturing and matching human motions, the grade of muscular stress, user-specific system parametrization, the way of delivering feedback and user-friendly interfaces for feedback and therapy evaluation. The presented system could contribute to a more efficient therapy, because it can supervise patients when the therapist is not present.

1 Introduction

In recent years, we have been facing the challenges that accompany the ageing of our society. One aim in future health care developments should be the preservation or even the improvement of every individual's health and quality of life by preventive and therapeutic measures. In the presented approach, we focus especially on the recovery of physical capabilities after a disease or an incident. In view of the

J. Richter (✉) · C. Wiede · A. Apitzsch · N. Nitzsche · C. Lösch ·
M. Weigert · T. Kronfeld · S. Weisleder · G. Hirtz
Technische Universität Chemnitz, Professorship Digital Signal Processing and Circuit Technology, Reichenhainer Str. 70, 09126 Chemnitz, Germany
e-mail: julia.richter@etit.tu-chemnitz.de

R. Wichert and B. Mand (eds.), *Ambient Assisted Living*,
Advanced Technologies and Societal Change,
DOI 10.1007/978-3-319-52322-4_8

decreasing number of personnel resources in the orthopaedic and neurological therapy sector, particularly in rural areas, it is sensible to assist patients and therapists by providing technical support systems in rehabilitation. By doing this, the rehabilitation process could significantly be improved.

Our current results from interviews with medical exercise therapists showed that presently, one therapist supervises up to fifteen patients, whereas the supervision of only one patient is rather rare (own data collection from interviews with therapists). Consequently, the therapist supervises several patients at the same time and therefore is not able to continuously supervise each patient's performance. The correct movement execution is, however, crucial for an effective and successful therapy. Different factors can influence the muscle activity (measured with an EMG) while performing exercises, e.g. the starting position [14], the movement amplitude [5, 9], the movement velocity and the exercise intensity [20]. In order to ensure that the intended muscles are active, it is essential to control the mentioned factors in medical exercise therapy. Moreover, sometimes patients tend to move other body parts as well although they should not move them. Every incorrect movement can lead to complications, because the exercise does not show the targeted effect or it can even have negative effects on the recovery.

Without a contemporary correction of errors, which are sneaking in during the therapy exercises, the overall therapy might not be successful and the patient could possibly not optimally recover from his or her disease. Therefore, the objective we pursue in our project "AssiSt" (Assisted Motion Control Using Smart Sensor Technology) is to emulate the therapist's knowledge and his or her visual perception in order to reproduce the therapist's real-time feedback. In that way, the exercise quality shall be secured, especially in the periods when the therapist cannot supervise the patient. The system is intended to be used in rehabilitation facilities where multiple users are performing therapy exercises. In future, an utilization at the patient's home would also be sensible. A higher quality in the rehabilitation process can lead to a faster patient's recovery and consequently to an earlier return to working life and to a reduced expenses in the health sector.

The paper is organized as follows: Sect. 2 gives an overview about the application and the benefits of real-time feedback in professional sports and provides insight into the current situation in the field of medical training therapy and the recent application of infrared thermography diagnostics. Moreover, recent developments with regard to therapy assessment and real-time feedback are presented. Based on the knowledge about the current status, we introduce an approach for a feedback-based therapy system in Sect. 3, which shall improve the current therapy process and assist both therapists and patients. Section 4 discusses the proposed approach and rise open questions that still have to be answered. Section 5 concludes the paper while giving an outlook on future work.

2 Related Work

2.1 Application of Feedback

In the past, researchers have been developing systems that can measure, process and visualize human motion data in the context of professional and competitive sports. This form of technique and condition training is especially used in diagnostics [16]. Another application field for feedback is balance and walking training for patient who have suffered an apoplectic stroke. Hamman et al. demonstrated that, for healthy probands, balance training with additional visual feedback results in a reduced swaying while walking [11]. Feedback also proved suitable for EMG in order to treat patients with reduced neuro-motoric control [4]. Consequently, recent evidence suggests that the application of real-time feedback to the user has a positive effect on the training results.

In the field of medical training therapy, however, real-time-based feedback can only scarcely be found in practice nowadays. Therefore, few studies have investigated the challenges and the effects of real-time feedback in medical training therapy. In the field of fitness training, several innovative products have already been employed. Although these products offer a wide range of acquirable data, they nevertheless show several disadvantages. The devices do not work contactless, which means that they have to be elaborately fixed on the human body. This may restrict the free movement of body parts and is not feasible in practical therapy applications. Furthermore, they are not capable to compare the performed movement against a reference movement. Besides that, the measured information is rather based on indirectly measured parameters in the training machine, e.g. the calculated torsional moment in the lever arm, than on directly measured parameters, e.g. the patient's skeletal joints and joint connections. On these grounds, the currently available products are not suitable for an application in the rehabilitation. Another aspect we address in our project is the documentation of the therapy. At present time, therapists have to create a handwritten therapy documentation normally containing the kind of exercises and the number of repetitions and the grade of difficulty. An automatic documentation with a detailed and clear presentation of occurred error patterns does not exist in practice yet.

2.2 Infrared Thermography

In medical diagnostics, surface temperature is increasingly exploited as an indicator for physical strain. Infrared thermography is used to determine temperature changes of musculature. In a recent study, Fröhlich et al. detected a relationship between the grade of physical strain and the heat development on the skin [6]. Bartuzi et al. verified statistically significant correlations between temperature and EMG parameters for the biceps brachii using infrared thermography as well [1]. In view of all that has been mentioned in these experimental studies, one may suppose that infrared thermography will also give valuable results in practical therapy environments.

2.3 E-rehabilitation and Motion Sequence Matching

E-rehabilitation and motion sequence matching is an upcoming field of research. Tak et al. [27] proposed a scheme for human abnormality detection for surveillance purposes. They used motion sequences as input data. Firstly, the human body was detected by applying a foreground-background segmentation on each image in the sequence. Secondly, their algorithm determined the contour around the detected foreground body. By calculating the distances between the geometric center point of the contour and all the points belonging to the contour, they obtained a distance curve for each image. In the next step, they performed a Fourier transformation for each distance curve. Each motion sequence was then described by a set of Fourier coefficients. Finally, they determined motion similarities by comparing a predefined human motion sequence with the motion that was performed by the user in real-time. For comparison, they used Dynamic Time Warping (DTW) and Dynamic Group Warping. For their approach they assumed a static background, which simplifies the part of human body detection in the image. However, if the background is changing or the person is interacting with the training machine, the task of human body detection will be more challenging. In such cases, simple foreground-background segmentation algorithms will prove error-prone.

Since the launch of the Microsoft Kinect in 2010, an increasing number of work employ this depth sensor with the according skeleton extraction algorithm developed by Shotton et al. [21, 22]. A lot of research focused on assessing human motions by means of evaluation skeletal information obtained by the Microsoft Kinect. Huang et al. [13] assessed dancing movements by comparing acquired skeleton data with a reference movement of a teacher. Similar to [27], they used DTW in order to match the sequences. Instead of silhouette curves, they processed the skeletal data by normalizing the joint positions with respect to the distance between spine and head. Thereupon, they calculated a body motion vector for each frame, which contains the relative movement of each body joint across two consecutive frames. Moreover, they determined the angle between two body motion vectors of the same joint, because it is less sensitive to variations in magnitudes of two consecutive body motion vectors. By taking this data into consideration, they matched the performed motion with the template sequence employing DTW. They obtained a final score for the performed dancing movement. For rhythm accuracy assessment, they performed a Fourier analysis on the magnitude of body motion vectors along the time axis. After the dancing movement had been performed, feedback was given after the performance in the form of the final scores a dancer achieved. Another Kinect-based approach that was designed for dancing assessment is the study of Kyan et al. [17]. They parsed ballet dance movement into a structured posture space using spherical self-organizing maps and cross-referenced against a library of gestural components performed by a teacher. The feedback was given through a CAVE virtual environment. Although this kind of feedback is very intuitive and tangible for the user, it is not feasible for practical therapy settings.

The presented algorithms originally were designed for assessing dancing movements. However, the principle work-flows could be reused in order to adapt it to therapy exercises in rehabilitation facilities.

Several recent studies have been investigating the employment of the Microsoft Kinect for therapy purposes. The majority of recent approaches utilize DTW and different variants of DTW. Several researchers employ fuzzy logic to categorize measurements according to similarity. Gal et al. [7] measured the angles between joints. Firstly, they created a physiological movement range for a certain patient by measuring starting and ending angles for each different exercise. Secondly, a fuzzy system evaluated delta values, which represent the distances between the physiological range and the actual range of motion. By defining angle limits, however, it is only possible to track the range of a motion. It is not possible to model the whole trajectory of a movement and consequently not suitable to assess whether the movements within the range have been correct. At this point, the work of Su et al. [26] introduced a home-based rehabilitation system that was designed to assess the trajectory and the speed of a therapy movement. They compared the performed exercise against a previously captured reference exercise from the same person. Su et al. decided to use DTW to compare the trajectories of the movements, because the patient may perform the exercises at a different duration. Since DTW allows an elastic deformation of the two time series, two sequences of different length can be easily compared by calculating the Euclidean distance between the sequence elements and by determining the optimal path that connects the starting and the ending element of the sequences. Su et al. used the 3-D joint coordinates of the wrist, elbow and shoulder joint as input data for the DTW algorithm. For the whole exercise sequence, they obtained distance measures for these three joints, which were the input for a neuro-fuzzy inference system. They defined several membership functions that assessed the trajectory in form of linguistic variables, i.e. dissimilar, similar or right. In the same manner, they evaluate the speed as bad, medium or right. A final performance evaluator combines the trajectory and speed evaluation in order to get a final assessment, i.e. bad, good or excellent. Although this final score includes the trajectory of the movement, it does not hold more detailed information about the trajectory. Moreover, in view of real-time feedback, the final score can only be presented to the patient after the exercise has been performed, e.g. in a playback. Kan et al. introduced Incremental Dynamic Time Warping (IDTW) in order to give a visual feedback after each acquired Kinect frame [15]. After skeleton normalization, they matched the partial user sequence with the complete reference sequence. By doing this, they could determine which joints differ from the reference motion. According to the IDTW distance, i.e. the grade of dissimilarity, they color-coded the respective joints.

The presented approaches are able to identify joints that do not fit the reference trajectory. However, an error classification that detects which possible incorrect movement has been performed, is not conducted. Together, these studies indicate that thus far there is still need to carry out research on solutions that provide detailed real-time feedback and that are feasible for a practical application in a rehabilitation facility or at home.

3 System Concept

This section introduces the proposed system with its single components. Figure 1 illustrates the rehabilitation system. The separate aspects of this system are presented in more detail in Sects. 3.1–3.4.

The system consists of a sensor that is able to extract the patient's skeleton and, based on this skeleton, measure the motion without attaching additional sensors to the patient. Furthermore, the temperature on the person's surface is monitored. This is realized with a sensor combination of a depth camera and a thermal camera. The final sensor will be installed in such a way, that the complete person is visible while performing the exercises. In contrast to state-of-the art motion capture systems, our system shall extract the skeleton from the images in a marker-less fashion. In case of current motion capture systems, a person has to put on several markers. However, this is not convenient for rehabilitation training. Possible disadvantaged of the motion capture with the Kinect sensor are discussed in Sect. 3.2. On the basis of the captured motion data, the system shall assess the performed movement and directly give suggestions for movement correction to the patient. In that way, the patient can immediately follow the suggestions how to correct a possibly incorrect movement. At this point, our research focuses on the individual parametrization of the system according to each patient's individual needs. We propose that the system should learn the correct movement while the patient is performing a certain exercise at the start of the therapy. This learning phase will be supervised by the therapist to ensure a correct movement execution, so that the system can store the correct movement.

In our project, we focus on exercises performed with a rope pull machine.

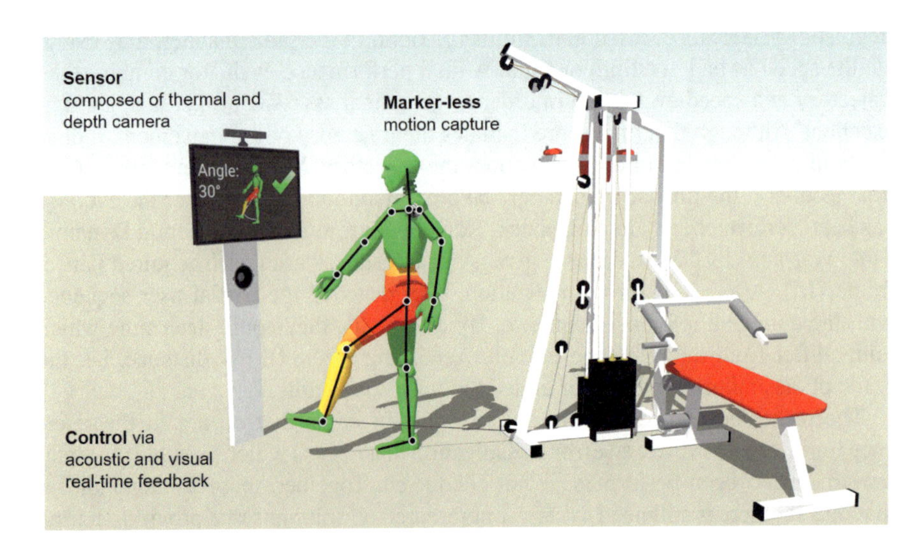

Fig. 1 Proposed rehabilitation system with sensing unit, i.e. depth and thermal sensor that shall capture the human motion in a marker- and contactless way. A feedback unit provides acoustic and visual instructions in order to control and, if necessary, correct the exercise performance

3.1 Sensors

The final sensor will be composed of a depth sensor, accompanied with a RGB camera of known extrinsic, and a thermal sensor. It will be part of further investigations what kind of depth sensor is the most suitable. Several criteria, such as field of view, image and depth resolution, density of the depth image, frame rate as well as synchronization aspects, have to be considered when deciding for a depth sensor. Depth sensors with the following principles are considered: time-of-flight, structured light and stereo sensors. The used thermal camera is a FLIR A35sc with a spatial resolution of 320×256 pixels and a thermal resolution of 0.05 K.

The two cameras have to be extrinsically calibrated. The precondition for the calibration is a calibration pattern that is visible for both kind of sensors, i.e. visible in RGB images and in the thermal image. Similar to [19], circular patterns that are heated can be used. A similar method would be to install a plate with cut-out circles in front of a monitor, so that the light of the screen illuminates the circles. Circular patterns have the advantage that the thermal expansion will not significantly influence the detected center points of the circles. Another solution is to floodlight a checkerboard with infra-red light, so that the different colors of the checkerboard will reflect the light differently.

3.2 Movement Control

Similar to the presented approaches in Sect. 2, one crucial component of our system will be the human pose estimation. After a robust skeleton extraction, the aim is to detect possible errors during the execution of the exercise and determine the necessary correction measures the user shall be given via the feedback units.

3.2.1 Human Pose Estimation

According to [18], human pose estimation is the process of inferring 2-D or 3-D human body part positions from still images or videos. Sigal et al. state that thus far, not any of these approaches can deal with unconstrained settings [24]. The following circumstances still challenge researchers working on unconstrained human skeleton extraction algorithms:

- variability in appearance, e.g. clothing
- variability in lighting conditions
- variability in human physique
- partial occlusions (self occlusions and occlusions caused by other objects)
- complexity of human skeletal structure

One of the most popular approaches in the field of human pose estimation is the method of Shotton et al. [21], which is running on the Microsoft Kinect. They demonstrated that their approach is able to accurately predict 3-D positions of human body joints in a single depth image without tracking algorithms or an underlying kinematic body model. Shotton et al. pointed out that traditional human body tracking algorithms, such as [2, 3, 8, 10, 25] exploit kinematic constraints and use temporal coherence from successive frames. However, according to [21], these tracking algorithms are prone to catastrophic loss of track if the tracked subjects are not regularly re-initialized. Although Shotton's original design was meant to include per-frame recovery, the final algorithm works without tracking, because their single-frame approach had proven to be sufficiently accurate.

Even the Kinect provides inaccurate results under difficult circumstances. In case of partial occlusions, i.e. self occlusions or occlusion caused by other objects, Shotton's approach revealed inaccuracies regarding the 3-D body joint proposals. Moreover, the algorithm is not able to infer a skeleton when a person is lying or viewed from the side. On the one hand, this is a problem for therapy exercises that require a lying position. On the other hand, it results in several restrictions with regard to the sensor's position on the training machine. When the patient has to frontally face the Kinect, then several exercises require a change of sensor position, which we would like to avoid in view of practical terms. Another issue we have to consider while comparing exercise movements is possible noise. As discussed in [13], the inferred skeletons can be noisy, i.e. there occur beating and jittering. For a possible solution see [13]. Further solutions for reducing noise and occlusion problems are discussed in [12, 23].

In order to increase the accuracy of skeletal joint proposals, we plan to invoke thermal image data and temporal information into the skeleton extraction process. Analogue to the depth features introduced by Shotton et al. [22], we want to construct thermal features, so that a classifier can be trained for different body parts. We plan to train the classifier in a similar way to Shotton's approach. By means of texture files, we want to generate synthetic training data by re-targeting human body models onto motion capture data.

3.2.2 Motion Sequence Matching

In order to give correction instructions directly after an incorrect movement, we plan to split the whole exercise into segments that shall be compared against a reference segment, e.g. one repetition of the movement. Since DTW proved suitable for comparing two sequences of different length, we intend to adapt this approach by evaluating DTW values for several joints.

In our project, we consider three different exercises: hip abduction, hip extension and hip flexion. Figure 2 shows examples for each exercise.

In order to efficiently design the assessment algorithm, we have to describe the different, possible errors in order to define the joints we want to consider in the DTW algorithm. This means that we only have to consider those joints that are relevant for

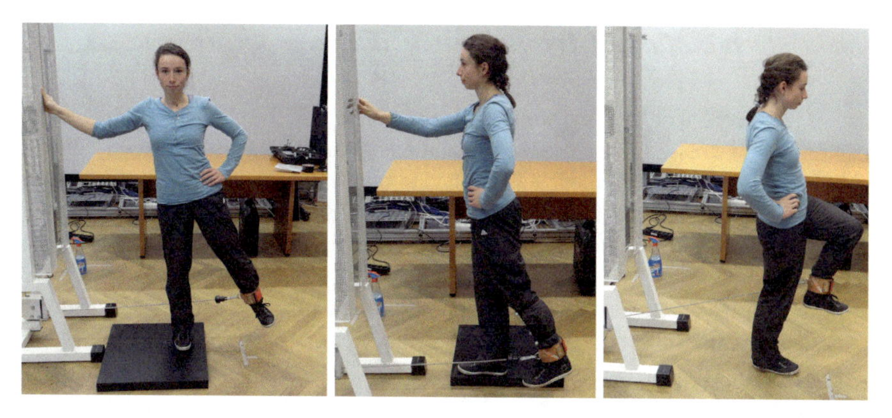

Fig. 2 Exercise examples for hip abduction, hip extension and hip flexion (from *left* to *right*)

the detection of possible errors. If the hand joints are not relevant for the exercise "Hip Abduction" with the right leg, then they can be discarded. A classifier shall be trained to recognize possible errors. Therefore, movements showing different types of errors are being recorded for different persons. In order to make the data rotation- and scale-invariant, the data is transformed in a local person's coordinate system and normalized according to the method described in [15].

In Table 1, descriptions of possible incorrect movements for "Hip Abduction" are listed. Please note that these errors can also occur as combinations. The descriptions for the other two exercises are formulated in an analog way. The described errors have to be identified by evaluating the skeletal joint information.

Table 1 Possible incorrect movements for hip abduction. Based on the descriptions, recordings showing incorrect movements can be generated and used for training a classifier that finally is able to detect these types of incorrect movements

Incorrect movement	Description
Moving upper body	The joints of the torso and head are moving although their position should not change
Different plane	The joints of the abducted leg are not moving in the plane that is spanned by the joints of the supporting leg and the straight torso
Rotated leg	The leg is rotated outwards or inwards, which results in a rotated position of the toe joint
High momentum	The movement is not performed smoothly, but with too much speed
High amplitude	The leg exceeds a certain angle when being abducted
Bent knee	The moving leg is not straight, but bend while performing the movement. The joints of the leg do not result in a straight line
Moving hip	The hip is not straight during the movement

3.3 Determination of Temperature Changes

One aspect in the feedback we want to give the patient is the change of temperature resulting from stressed muscles. We intend to give the feedback by color-coding the relevant body parts on the feedback display. In this way, the patient can see whether the exercise is performed in an efficient way, because the muscles involved in the exercise will turn from blue, corresponding to cold, to red, corresponding to warm, for example.

In this project, we mean to quantize the thermal expansion on the skin by evaluating the acquired thermal image data. Furthermore, a mapping to different body parts shall be realized. This goes hand in hand with body parts that have been determined during the pose estimation and the extrinsic calibration: for every pixel that shares the same body part label, the thermal image data of the calibrated, underlying thermal image can be evaluated by detecting changes over a certain period of time.

3.4 Real-Time Feedback and Documentation

The real-time feedback can prevent the user to adapt incorrect movements. The feedback shall be given in visual or acoustic form, e.g. via a display, where correction suggestions are presented, or via an additional loudspeaker that simultaneously gives the patient instructions. Next to the motion control feedback, the determined thermal changes on the surface of the skin shall be visualized using a color code.

Beside the given feedback, the system additionally shall document the exercise type, the number of repetitions and the quality of each performed exercise. The data shall be stored in a database, so that the therapist and the patient can evaluate the exercises afterwards, e.g. via a web front-end. Moreover, the automatic documentation supports the therapist, so that he or she is not burdened with time-consuming documentation work.

3.5 Interviews

During our project, we have developed different kinds of interviews for different target groups in order to determine special needs that therapists identified during their work and to get an insight into the patient's therapy experiences.

The requirement analysis is designed in the form of oral interviews with therapists. During the interviews, therapists are asked about the quality of therapies, the relationship to their patients, the documentation process and frequently occurring errors during the exercises. Finally, they shall answer questions related to benefits and challenges they anticipate when using the proposed assistance system.

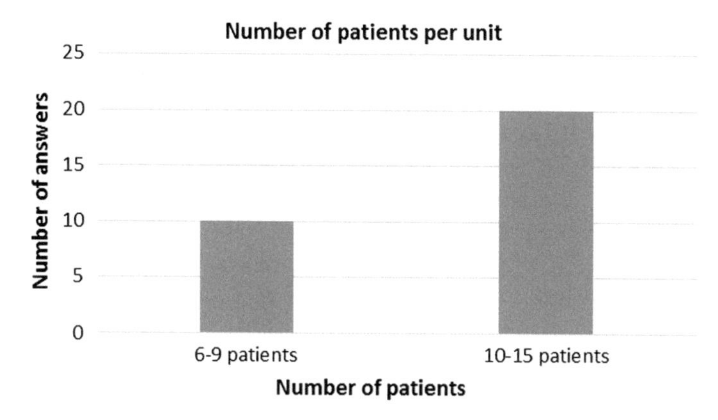

Fig. 3 Number of patients that have to be supervised during a session/unit: the majority of the interviewed experts says that 10–15 patients have to be supervised during a unit. For this question, only 30 experts are considered: One of the 31 interviewed experts is a physiotherapist. This group of experts generally supervise only one patient at a time

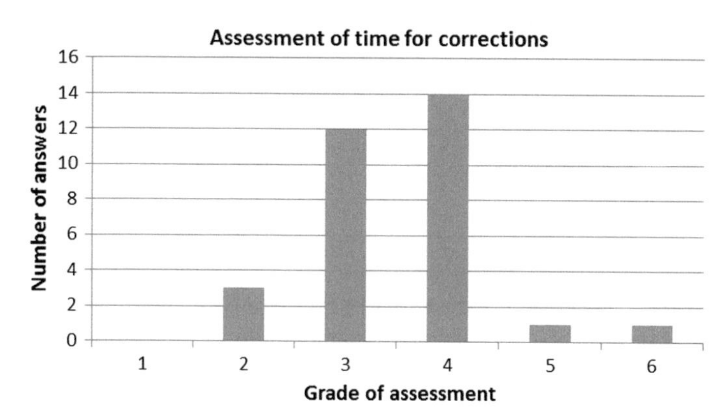

Fig. 4 Assessment of the time available for corrections: Experts were asked to assess this time by using grades from 1 to 6. The grade 1 indicates that they regard the time as perfectly sufficient, whereas the grade 6 means that the time is extremely insufficient

Our recent results are based on interviews with 31 sports- and physiotherapists from ten different rehabilitation facilities. The outcome of the interviews not only underline the need of technical assistance during therapy sessions, but they also indicate the requirements that have to be considered while designing such an assistance system.

Figures 3 and 4 illustrate the results regarding the number of patients that have to be supervised per session/unit and the experts' assessments about the time they have at their disposal to correct a patient's movement. The chart in Fig. 3 gives evidence that the number of patients a therapist has to supervise at the same time ranges from six to fifteen. This number is certainly too high if the movement quality shall be continuously ensured for all patients. The given answers in Fig. 4 clearly show that the therapists assess the available time they have to correct the patients' movements as insufficient.

For the patients, we created questionnaires asking questions with regard to the frequency, the duration and the experience with machine training. Moreover we want to determine how patients estimate their current training situation, including the quantity and quality of the supervision and feedback given by the therapist, and whether they are open-minded to a technically assisted training and where the patient anticipates possible challenges.

4 Discussion

This section will discuss the main challenges that might be faced during the development of an user-friendly assistance system that shall enable the patient to conduct a successful therapy, either at home or in a rehabilitation facility.

One of the questions that has to be answered first is the position of the sensor unit and the feedback display. On the one hand this aspect is important with regard to the algorithm development, such as the models used for human skeleton extraction. On the other hand, the position has influence on the practical applicability. The user should not be strained with changing the sensor's position, so that it has an optimal view on the patient. The sensor should rather be fixed at some position where the view is optimal for every exercise. Accordingly, the algorithms should be designed in a view-invariant way, so that a skeleton can be extracted from both frontal view and side view, for example. The position of the feedback display, however, should be variable, so that the user can change the position to his or her special needs. Taking this into consideration, the feedback display and the sensor unit should be designed as separate components.

In a next step, the accuracy of the calibration has to be determined. The accuracy should be of high quality for an exact mapping of depth and thermal images. There is also the possibility to map data on high-level, e.g. by finding body part positions in both thermal and depth image and map on this high-level description. Moreover, the different sensors have to be synchronized.

The skeleton extraction algorithm has to be analyzed with respect to joint position accuracy and invariance regarding the parameters mentioned in Sect. 3.2.1. Another aspect we have to investigate is the degree of temperature changes in thermal image while performing the exercise.

Finally, we attach high importance to usability. We want to avoid manual parameter input by the therapist while adapting the system to the individual user. Moreover, taking the future users' opinion into consideration, the kind of feedback should be intuitive and motivating.

5 Conclusions

In this investigation, we gave an overview about recent research that has been carried out in the field of motion control and real-time feedback in therapy. One of the most significant findings to emerge from this study is the identification of issues that still require further research, i.e. accurate pose estimation, motion sequence matching and the design of a real-time feedback. We therefore presented a concept for assisted motion control in therapy environments and discussed several aspects that have to be taken into consideration when designing such a practical system. Up to now, we considered three different exercises for a rope pull machine. Further work needs to be done to expand the approach to further exercises and other training machines. Moreover, it would be sensible to design the system for home-applications. Especially when patients are training alone, the motivation and enjoyment while doing the exercises should be stimulated.

With regard to demographic developments, the quality of rehabilitation has to be ensured. The proposed approach can contribute to improve the exercise quality in phases when the therapist is not present to supervise the patient. By identifying incorrect movements and by immediately giving feedback, we moreover expect a motivation effect for the patient. Not only would this be of high benefit for the patient's recovery, but it could also support therapists to supervise an increasing number of patients.

Acknowledgements This project is funded by the European Social Fund (ESF).

References

1. Bartuzi, P., Roman-Liu, D., Wiśniewski, T.: The influence of fatigue on muscle temperature. Int. J. Occup. Saf. Ergon. **18**(2), 233–243 (2012)
2. Bregler, C., Malik, J.: Tracking people with twists and exponential maps. In: IEEE Computer Society Conference on Computer Vision and Pattern Recognition, 1998. Proceedings, pp. 8–15. IEEE (1998)
3. Brubaker, M.A., Fleet, D.J., Hertzmann, A.: Physics-based person tracking using the anthropomorphic walker. Int. J. Comput. Vis. **87**(1–2), 140–155 (2010)
4. Brudny, J., Korein, J., Grynbaum, B.B., Friedmann, L.W., Weinstein, S., Sachs-Frankel, G., Belandres, P.V.: EMG feedback therapy: review of treatment of 114 patients. Arch. Phys. Med. Rehabil. **57**(2), 55–61 (1976)
5. Caterisano, A., Moss, R.E., Pellinger, T.K., Woodruff, K., Lewis, V.C., Booth, W., Khadra, T.: The effect of back squat depth on the EMG activity of 4 superficial hip and thigh muscles. J. Strength Cond. Res. **16**(3), 428–432 (2002)
6. Fröhlich, M., Ludwig, O., Zeller, P., Felder, H.: Changes in skin surface temperature after a 10-minute warm-up on a bike ergometer. Int. J. Kinesiol. Sports Sci. **3**(3), 13–17 (2015)
7. Gal, N., Andrei, D., Nemeş, D., Nădăşan, E., Stoicu-Tivadar, V.: A kinect based intelligent e-rehabilitation system in physical therapy. In: Digital Healthcare Empowering Europeans, pp. 489–493 (2015)

8. Ganapathi, V., Plagemann, C., Koller, D., Thrun, S.: Real time motion capture using a single time-of-flight camera. In: IEEE Conference on Computer Vision and Pattern Recognition (CVPR), 2010, pp. 755–762. IEEE (2010)
9. Gorsuch, J., Long, J., Miller, K., Primeau, K., Rutledge, S., Sossong, A., Durocher, J.J.: The effect of squat depth on multiarticular muscle activation in collegiate cross-country runners. J. Strength Cond. Res. **27**(9):2619–2625 (2013)
10. Grest, D., Woetzel, J., Koch, R.: Nonlinear body pose estimation from depth images. In: Pattern Recognition, pp. 285–292. Springer (2005)
11. Hamman, R.G., Mekjavic, I., Mallinson, A.I., Longridge, N.S.: Training effects during repeated therapy sessions of balance training using visual feedback. Arch. Phys. Med. Rehabil. **72**(8), 738–744 (1992)
12. Huang, H.Y., Chang, S.H.: A skeleton-occluded repair method from kinect. In: International Symposium on Computer, Consumer and Control (IS3C), 2014, pp. 264–267. IEEE (2014)
13. Huang, T.-C., Cheng, Y.-C., Chiang, C.-C.: Automatic dancing assessment using kinect. In: Advances in Intelligent Systems and Applications, Vol. 2, pp. 511–520. Springer (2013)
14. Kang, S.-Y., Choung, S.-D., Jeon, H.-S.: Modifying the hip abduction angle during bridging exercise can facilitate gluteus maximus activity. Man. Ther. (2016)
15. Khan, N.M., Lin, S., Guan, L., Guo, B.: A visual evaluation framework for in-home physical rehabilitation. In: 2014 IEEE International Symposium on Multimedia (ISM), pp. 237–240. IEEE (2014)
16. Krug, J., Herrmann, H., Naundorf, F., Panzer, S., Wagner, K.: Messplatztraining: Konzepte, Entwicklungsstand und Ausblick. Messplatztraining, pp. 13–27 (2004)
17. Kyan, M., Sun, G., Li, H., Zhong, L., Muneesawang, P., Dong, N., Elder, B., Guan, L.: An approach to ballet dance training through MS kinect and visualization in a CAVE virtual reality environment. ACM Trans. Intell. Syst. Technol. **6**(2), 23:1–23:37 (2015)
18. Liu, Z., Zhu, J., Bu, J., Chen, C.: A survey of human pose estimation: the body parts parsing based methods. J. Vis. Commun. Image Rep. **32**, 10–19 (2015)
19. Nakagawa, W., Matsumoto, K., de Sorbier, F., Sugimoto, M., Saito, H., Senda, S., Shibata, T., Iketani, A.: Visualization of temperature change using RGB-D camera and thermal camera. In: Computer Vision-ECCV 2014 Workshops, pp. 386–400. Springer (2014)
20. Sakamoto, A., Sinclair, P.J.: Muscle activations under varying lifting speeds and intensities during bench press. Eur. J. Appl. Physiol. **112**(3), 1015–1025 (2012)
21. Shotton, J., Girshick, R., Fitzgibbon, A., Sharp, T., Cook, M., Finocchio, M., Moore, R., Kohli, P., Criminisi, A., Kipman, A., et al.: Efficient human pose estimation from single depth images. IEEE Trans. Pattern Anal. Mach. Intell. **35**(12), 2821–2840 (2013)
22. Shotton, J., Sharp, T., Kipman, A., Fitzgibbon, A., Finocchio, M., Blake, A., Cook, M., Moore, R.: Real-time human pose recognition in parts from single depth images. Commun. ACM **56**(1), 116–124 (2013)
23. Shum, H.P.H., Ho, E.S.L., Jiang, Y., Takagi, S.: Real-time posture reconstruction for microsoft kinect. IEEE Trans. Cybern. **43**(5), 1357–1369 (2013)
24. Sigal, L.: Human pose estimation. In: *Computer Vision*, pp. 362–370. Springer (2014)
25. Sigal, L., Bhatia, S., Roth, S., Black, M.J., Isard, M.: Tracking loose-limbed people. In *Proceedings of the 2004 IEEE Computer Society Conference on Computer Vision and Pattern Recognition, 2004. CVPR 2004*, vol. 1, pp. I–421. IEEE (2004)
26. Su, C.-J., Chiang, C.-Y., Huang, J.-Y.: Kinect-enabled home-based rehabilitation system using dynamic time warping and fuzzy logic. Appl. Soft Comput. **22**, 652–666 (2014)
27. Tak, Y.-S., Rho, S., Hwang, E.: Motion sequence-based human abnormality detection scheme for smart spaces. Wirel. Pers. Commun. **60**(3), 507–519 (2011)

eNav: A Suitable Navigation System for the Disabled

Dženan Džafić, Pierre Schoonbrood, Dominik Franke and Stefan Kowalewski

Abstract Navigation for disabled people in wheelchairs is a huge challenge: stairs and curbstones can already hinder a wheelchair to pass the route. The availability of affordable increasingly powerful smartphones and Internet allows for smart applications, which can help wheelchairs users to find a valid and accessible route. In this paper eNav is presented, which is such an application. It brings together different features to help routing for wheelchairs. eNav links big data from areas of maps, incline and different disability databases to compute the most energy-efficient route between two points. The complex and cumulated data is presented to the disabled in a very simple and understandable manner via desktop PCs or smartphones. By using crowd sourcing technology, eNav users have the possibility to increase the map data used by eNav. An evaluation between the energy-efficient and the shortest route shows that in 41% of all tested routes, the efficient route uses less energy than the shortest.

Keywords Disabled · Wheelchair · Navigation · eNav · Energy efficiency · Crowdsourcing · OSM · POI · Accessibility

D. Džafić (✉) · P. Schoonbrood (✉) · D. Franke (✉) · S. Kowalewski (✉)
RWTH Aachen Embedded Software, Ahornstraße 55, 52074 Aachen, Germany
e-mail: dzafic@embedded.rwth-aachen.de

P. Schoonbrood
e-mail: schoonbrood@embedded.rwth-aachen.de

D. Franke
e-mail: franke@embedded.rwth-aachen.de

S. Kowalewski
e-mail: kowalewski@embedded.rwth-aachen.de

R. Wichert and B. Mand (eds.), *Ambient Assisted Living*,
Advanced Technologies and Societal Change,
DOI 10.1007/978-3-319-52322-4_9

1 Introduction

There are 7.5 million people living in Germany which are severely disabled. This corresponds to 9.4% of Germany's total population [1]. This community should not be ignored during the development of navigation systems.

A specific routing system requires more than accessible routing. Electric wheelchair user not only have problems judging whether a route is barrier free or not, they also need to determine whether the battery level of their wheelchair suffices to bring them to their destination and, even more importantly, back home.

Battery indicators of electric wheelchairs are generally not reliable. On one hand, they behave in the same way as battery indicators of old cellphones. The battery level is indicted as full most of the discharge time. If the battery is discharged to a certain level, the battery indicator quickly depicts the battery as depleted. On the other hand, even if the battery indicator would depict depletion linearly, it is hard to determine the reach of a certain battery charge level. This is caused by the strong dependency between the incline and the energy consumption of electric wheelchairs [2].

To tackle these challenges, the map-data used in common navigation systems needs to be enriched with relevant data like altitude and surface information.

In this work, the eNav navigation system is presented. It proposes smart and innovative solutions to these challenges. It is basically founded on a wide collection of different maps extended with detailed information about the nature and the quality of the roads which hinder wheelchair user. To collect the necessary map-data for this task, several distinctive sources and methods are used.

In the next section, basic notions and definitions are given. Afterwards, the architecture of eNav is discussed, followed by a detailed explanation of the components of the architecture as well as the methods used in eNav for computing the energy efficient route. Finally, the profitability of the most energy efficient route is evaluated in contrast to the shortest route.

2 Preliminaries

This section gives a short definition of accessibility with regard to handicapped people. It furthermore describes the nature and reveals the source of road or rural maps upon which our special navigation system eNav is build. In addition, it introduces the A^*-algorithm adopted for the eNav distance computation. It finally concludes with an overview of the crowd sourcing concept which is used to enhance data fed into eNav.

2.1 Accessibility

To guarantee for equal rights between physically handicapped and non-handicapped people, the German law defines the notion of accessibility as follows:

> Building works and other constructions are accessible [...], if they are usable and accessible by the disabled in the usual way, without any unusual complications principally unaided. (BGG Sect. 4)

This legislation consequently imposes an environmental infrastructure allowing disabled people to use it at least with the same effort as persons without handicaps would do. This work focuses essentially on accessibility notion related to mobility impaired people.

2.2 Map Sources

To enable accessible routing, eNav uses a three-dimensional map which is enriched with additional information from several distinct sources. OpenStreetMap (OSM) is used as a main source of base maps. These maps are extended with laser scan data and surface type information, which is provided by the municipality of Aachen.

2.2.1 OpenStreetMap (OSM)

The OSM platform was the best free source found to acquire maps for eNav. According to [3], no other open source project exists with a comparable world map data density and reliability like OSM. Additionally, OSM provides a good basis, to extent its data, available in XML-Format, with project specific data. There are tags for this purpose, which describe a property and are stored as a key-value pair in the OSM-files. For three-dimensional routing, especially the tags with the key altitude and elevation are of interest. These tags store height information of nodes. Considering eNav and its routing, several other tags like "wheelchair", "step", etc. are of interest as well.

2.2.2 Laser Scan Data

The height information indispensable for eNav is only sparsely available in the OSM-maps [4]. The integration of this information into the OSM-maps is of decisive importance not only to eNav but also to other projects, like a simple 3D illustration of OSM-maps [5]. This crucial height information with which the OSM-maps are enriched is provided by laser scan data.

This data is acquired during an earth surface scanning done by an airplane using laser tracking with a frequency of approximately 200 kHz. The laser pulses send from the airplane are reflected on the earth's surface and then captured again by the airplane sensors. From the resulting elapsed time, the relative distance between the laser scanner on the airplane and the earth's surface is calculated. Using other sensors on the airplane, like GPS and IMU, the relative distance to the absolute height of earth's surface is recalculated.

During scanning of large areas, the airplane flies striped patterns possessing a width of several hundred meters. In eNav, a digital surface model with a 1 m raster (DOM1) is used. DOM1 has an accuracy of ±20 cm, which suffices for the route calculation used in eNav [6]. The district government of Cologne provided the laser scan data of Aachen for the eNav project, which must be used solely for research purposes.

2.3 A^*-Algorithm

The A^*-algorithm calcul ates the shortest path from A to B. It adopts a similar approach to that of the Dijkstra's [7] algorithm. However, A^* uses a heuristic in addition to path weights to conduct a targeted search and reduce thereby the execution time. An example of such a heuristic is the linear distance between intermediate nodes and the target [7, 8].

2.4 Crowdsourcing

Crowdsourcing is a web-based business model originally proposed by Jeff Howe in 2006 [9]. It is the action of an institution to outsource a task, usually done by an employee, to a network commonly with a large user base in an open tender. On the one hand, the task can be designed to gather new knowledge contributing to the solution of a problem; on the other hand, it can be a contribution to the marketing or the configuration of a product [10].

Part of crowd sourcing is crowd testing. The latter notion originates from software development field. This special test finds software faults by daily monitoring and analyzing the activity of a large user base [11]. In eNav, this technique is not used to detect software faults, but rather to discover inaccuracies in the data used by the software.

3 Related Work

In this section, recent projects are listed, which are based on approaches similar to the method adopted for eNav. The projects mPass [12], Easywheel [13], and Path2.0 [14] are excluded from this overview because of their similarities to the OpenRouteService [15] project. Projects like inDAgo [16] incorporate accessible routing as well, but their target group are the elderly.

3.1 OpenRouteService

OpenRouteService (ORS) is an open source route planner based on OSM-maps. The University of Bonn is the original developer of this project, which is currently under active development at the University of Heidelberg. Since the OSM-data used is popular and widespread, the resulting routes are highly reliable. The modes that ORS supports are car, bicycle and pedestrian, which generally suites only common users. However, incline information is only sparsely available. Since ORS applies the classical A^* search algorithm, incline information is not taken into account during the route computation. That results in distance and time being the unique determining factors for such a computation. For the disabled, a new computation approach *rollstohlrouting.de* is developed. This approach involves the sparsely available incline information, the level of a curbside, the pavement and the width of a carriageway as important factors in route planning.

The ORS service is accessible for users via the *openrouteservice.org*. In addition, other application and clients can use the service via an API. As usual, the user can decide to travel the fastest or the shortest route. This choice influences the calculation. The user is supported by an adequate route description including an output of the route on a map.

3.2 Wheelmap

The founder of *Wheelmap* is the registered association *Sozialhelden e.V*, which is headed by Raul Krauthausen. The main contribution of *Wheelmap* [17] is the ability to tag points of interest (POI) whether they are wheelchair accessible or not and additionally, whether a wheelchair accessible toilet is available or not. In contrast to eNav, *Wheelmap* does not provide any routing functions. The API of *Wheelmap* is used to integrate POI into the eNav navigation system.

4 eNav

eNav is a navigation system for the disabled, which enables them besides accessible routing, to choose a route which improves driving comfort. Additionally, eNav grants electric wheelchair users a better overview of the reach of the battery pack. Usually, battery indication on electric wheelchairs is as inaccurate as those of cellphones. During most of time, the battery is indicated as being full. After some usage time, the battery is quickly indicated as being depleted. An electric wheelchair user can easily make a miscalculation because of this problem. Although he can easily reach his destination, he may not be to go back home.

eNav provides the possibility to calculate the shortest route to a destination, as well as the most energy efficient. Moreover, it presents a course overview about how much of a charged battery is used during the planned trip. Furthermore, by using the *SpiderWebGraph* algorithm [18], eNav is one of the first navigation systems allowing direct routing across flat areas like squares.

A complete software architecture description of eNav is given is the next section followed by a detailed description of single components.

4.1 Architecture

The architecture of eNav is based on the classical client-server architecture. That means the user queries the client for a route. This query is then sent as request to a server. The server in its turn calculates the routes and sends them back as answer to the requesting client, where they are presented to the querying user.

The map-data used for the route calculation is gathered from several data sources, as seen in Fig. 1. Furthermore, the map-data is constantly updated with the help of crowdsourcing.

The next section presents the detailed concept of the client part architecture.

4.1.1 Client

Currently, there is a web-based client for desktops and one for smartphones (Fig. 1). Both enable the user to enter additional parameters, as can be seen in Fig. 2 on the left side. A part of these parameters is reserved for properties of the electric wheelchair with which the user wants to travel. These parameters enable, besides the calculation of the shortest route, the calculation of the most energy efficient route.

Figure 2 in the middle shows the choice the user can have, when the shortest and the most energy efficient route differ. If the user selects the energy efficient route, then the distance is increased by 230 m, but the energy consumption is reduced by

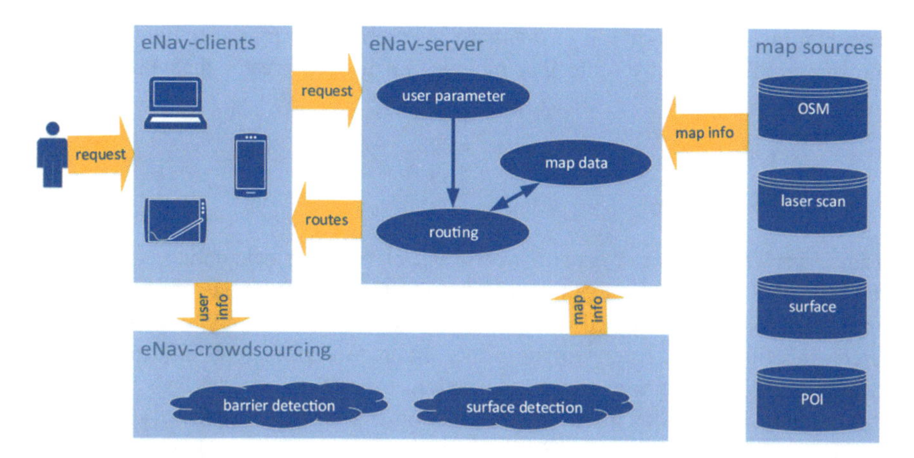

Fig. 1 eNav architecture and its components

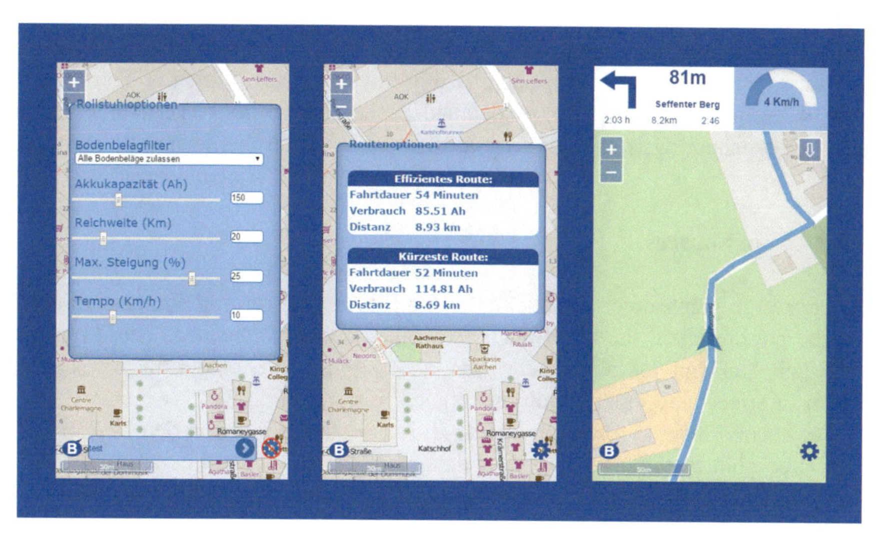

Fig. 2 The user interface of eNav. (1) Routing options. (2) Choice between shortest and efficient route. (3) eNav output during navigation

ca. 25% in contrast to the shortest. After selecting one of the two choices, navigation is started and the screen shown in Fig. 2 on the right can be seen.

The surface type can also be added as an optional parameter for planning the route. A wheelchair user can, for example, choose to avoid cobblestone. This will increase energy consumption, in addition to a possible increase of driving comfort.

The parameter incline can be explained twofold. For electric wheelchair users, this parameter can be translated to the maximal incline the wheelchair can cope with. For hand-operated wheelchair users, this parameter can be interpreted as the maximal incline the user can or wants to overcome. During the route calculation, the server rules out any route that includes an incline greater than the maximal incline set by the user.

4.1.2 Server

The server (Fig. 1) takes care of every aspect of route calculation. Clients can request route calculations using a REST API. These calculations are based on an especially for eNav designed directed graph object, in which the edges are annotated with their energy consumption in addition to surface type information. For calculation of the edges energy consumption, every node is enriched with height information. Information about this calculation can be found in Sect. 4.2.

The route calculation is conducted with an especially for eNav modified A^*-algorithm, which requires a heuristic suitable for considering energy efficiency. A more detailed explanation of the modification of this algorithm can be found in [19].

The information required for the graph object used for routing, is collected from several sources. Several sources apart from crowdsourcing, which is described in Sect. 4.1.4, are explained in more detailed in the next section. The data collected using crowdsourcing is send directly to the server, which then evaluates them.

4.1.3 Map Sources

To create the graph object mentioned in the last section, several sources are used. On the bottom layer, the OSM-maps (see Sect. 2.2.1) are situated. These provide eNav with information about addresses, accessibility, street- and track positions, length of streets and other conditions related to traffic (Fig. 3).

The second layer is formed by the laser scan data described in Sect. 2.2.2. This data is combined with the laser scan data from the first layer in such a way that every node from the bottom layer is matched to a data point of the laser scan data layer using clustering. This combination produces a three-dimensional map and allows the calculation of the incline information at every edge.

After the creation of a three-dimensional map, the third layer, which consists of the surface type information delivered by the municipality of Aachen, can be stacked on top of it. This data entails information about every track on a street,

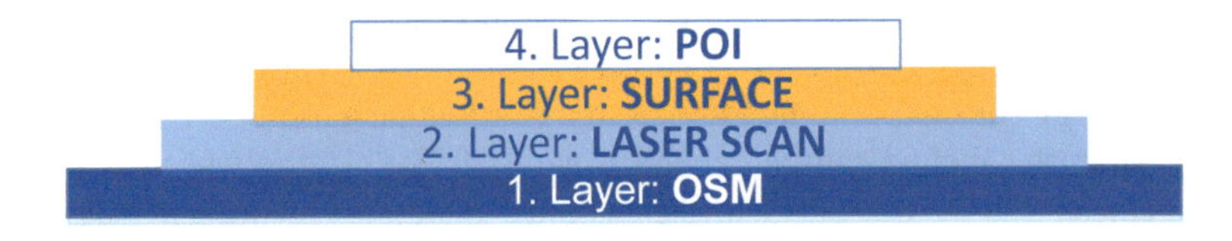

Fig. 3 Map-layers

which requires matching of two surface types at one edge. This constitutes a challenging task. In fact, consider an edge with a footway and a bicycle tag on the first level. On the third level, the footway has a different surface type than the bicycle track. Currently, only one surface type can be matched with an edge. As a solution, a best of strategy in terms of comfort is used, as a wheelchair user may use both tracks.

The last layer is comprised out of the *Wheelmap* database (Sect. 3.2), which supplies eNav with information about the accessibility of POI.

4.1.4 Crowdsourcing

Crowdsourcing serves eNav on three aspects of data refinement:

1. **Barrier detection**:
   In Sect. 2.2.1 it is stated that the quality of data in OSM is very high, nevertheless this data is incomplete in regard to several aspects. As a consequence, several steps and other barriers exist, which are not marked as such in OSM. To make this date more complete, the eNav GUI enables users to report barriers in the way illustrated in Fig. 4. Thereby a distinction is made between permanent and temporal barriers. A permanent barrier definitely blocks the passage for a wheelchair user, as for example stairs. Temporal barrier block the passage for a limited time. A car parked on the sidewalk would be an example of such a barrier. This differentiation is made to hinder temporal barriers to block edges for a long time. If temporal barriers, like roadworks, are reported, then the concerning edge is blocked only for a predefined period of time.
   In addition to manual barrier detection, crowd-testing is used to scan barriers automatically. The client is able to detect where and when users deviate from the proposed route and then reports this information to the server. Whenever the server detects a large number of such deviations confined in a certain spot, it alarms an Administrator group. The attending administrator then decides if the concerning edge is excluded from routing or not. Over time, the reliability of the data in terms of accessibility is improved by this procedure.
2. **Surface detection**:
   The quality of surface type data delivered by the municipality of Aachen is very good. In rural areas, however, this information is not as detailed as in urban areas or in some cases not available at all. To ensure a high quality of surface type information outside the district of Aachen, eNav uses a surface detector. This detector measures the linear acceleration along the z-axis using a smartphone. The acceleration is then used to differentiate between cobblestones and asphalt. Several other projects take a similar approach for collecting this type of data [20]. The user constantly stores the linear acceleration in the z-axis during the trip. By comparing the acceleration of edges from which the surface type is known beforehand and of those which are unknown, the surface type of the

Fig. 4 The barrier detection interface

unknown area can be deduced. This enabled automatic calibration of the detector and that potentially differentiates eNav from other projects.

A distinction is made between surface types which have a noticeable impact on driving comfort. As explained in Sect. 4.1.3, the problem of matching surface type of different tracks on one edge is an emerging problem, which is actually well solved with a "best of"-approach.

3. **The POI**:
   The POI are only indirectly integrated into eNav by outsourcing them to *wheelmap.org*, managed by *the Sozialhelden e.V.* (Sect. 3.2). Thereby the information entered by eNav users concerning the POI are relegated to *Wheelmap.org*.

4.2 Energy Consumption Function

In this section the consumption function used by the modified A^*-algorithm to calculate the most energy efficient route is explained. This function takes two influencing factors into consideration. First, the incline influencing factor will be described. Afterwards, surface types and their friction coefficients are discussed.

4.2.1 Incline

To examine the influence incline has on energy consumption; sensors have been installed on a wheelchair, which constantly measure energy consumption during driving [21]. Several test drives have been conducted with a C500 electric wheelchair of the company Permobil, to gather empirical data for several different slopes. This is data is used to deduce following consumption function:

$$\text{consumption} := \text{normal consumption} * \text{distance} * 1.16^{\text{incline}}$$

The normal consumption denotes the energy usage consumption while driving on an even surface, which is calculated by dividing the full load capacity of the battery by the lower limit of the maximum reach of the electric wheelchair. As it can be seen, the energy consumption has an exponential behavior in regard to incline, which can be explained by the discharge time of a battery. Figure 5 shows this exponential behavior. A more detailed description can be found in [2].

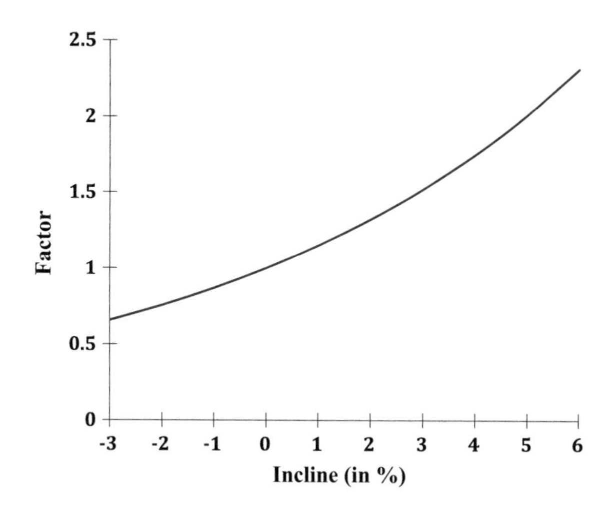

Fig. 5 Relation between energy consumption and incline

4.2.2 Friction Coefficient

To take the friction effect of surfaces into consideration during the computation, a friction coefficient should be assigned to each surface type. These coefficients have been determined with the help of the research group of *Professor Oeser* from the *Institute of road engineering of the RWTH Aachen*. A detailed description and complete table of friction coefficients can be found in [22, 23]. The consumption function is expanded to incorporate these coefficients:

$$\text{consumption} := \text{normal consumption} * \text{distance} * 1.16^{\text{incline}} * \text{friction coefficient}.$$

The factor for the incline has been determined on asphalt, which is used as the reference surface type and therefore assigned the factor 1 as a friction coefficient. Another coefficient is for example cobblestone coefficient with the factor 1.05. However, the surface type has a minimal effect on the overall energy consumption when compared with the incline. The modification of the heuristic of the A^*-algorithm to incorporate this consumption function is discussed in [19].

5 Evaluation

In this section presents an eNav evaluation based on different tests. During these the shortest route is compared against the most energy efficient one. During this evaluation the focus set on the energy consumption reduction the efficient route yields in contrast to the shortest one. A more specific test is not conducted yet. Nevertheless, indicators exist which show that the profitability of the most energy efficient route is clearly higher than the profitability of the shortest one.

5.1 Test Procedure

For the test, 100.000 routes have been calculated with randomly generated start points and destinations. All these points are situated within Aachen city. The deepest point is at 125 MASL and the highest at 410 MASL. For every 100.000 routes, both the shortest as the most energy efficient route is calculated. For data evaluation, the start point and destination, as well as the distance and the energy consumption for both routes is recorded. Using a *PostgreSQL* database, this data is evaluated. The consumption function including both incline and surface type (Sect. 4.2) is used for the calculation of the most energy efficient route.

5.2 Shortest Versus Efficient Route

First, the number of routes where the energy consumption of the efficient route is lower than the shortest route is counted. Figure 6 shows the percentage of efficient routes which consume less energy than the shortest route. In 41% of the cases energy can be saved by choosing the efficient one. In the other 59%, the energy consumption of the efficient one equals that of the shortest route.

To better judge the quality of these measurements, it should be mentioned that urban area of Aachen has up to 285 m difference in altitude. In more flat areas the percentage differs. By calculating the average of the energy saving in percentage, an increase of efficiency of 2% can be observed. The main cause of this value is that the major part of shortest and efficient routes only differ 1% in terms of energy efficiency. A switch between the sides of the street is responsible for this phenomenon.

Figure 7 shows the relation between the efficient route and the shortest route using a filter. The filter sorts out every case where the efficient route only differs up to 1% in energy efficiency in comparison with the shortest one. This filter is necessary, because the occurrence of this 1% difference is very high and therefore can distort the result.

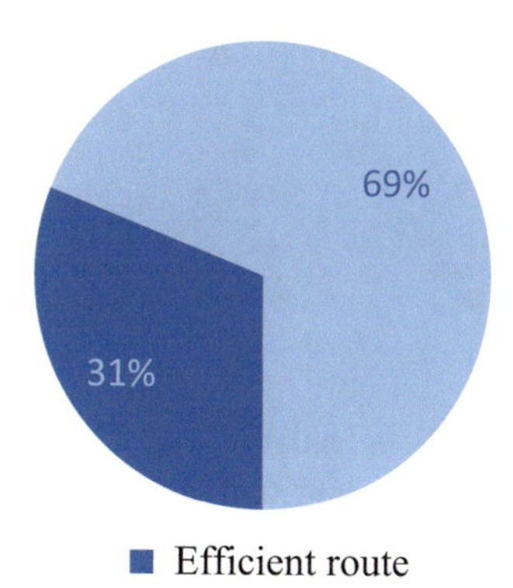

Fig. 6 Route comparison of test cases

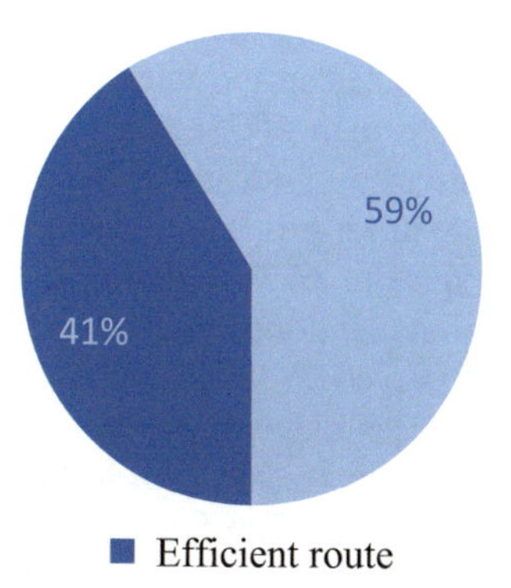

Fig. 7 Route comparison of test cases including <1% cases

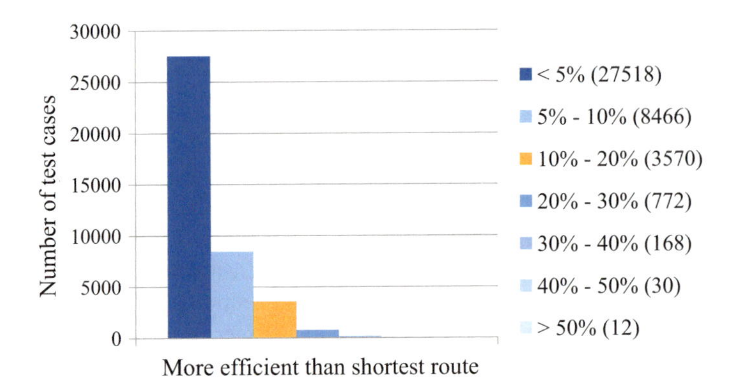

Fig. 8 Distribution of the efficient routes

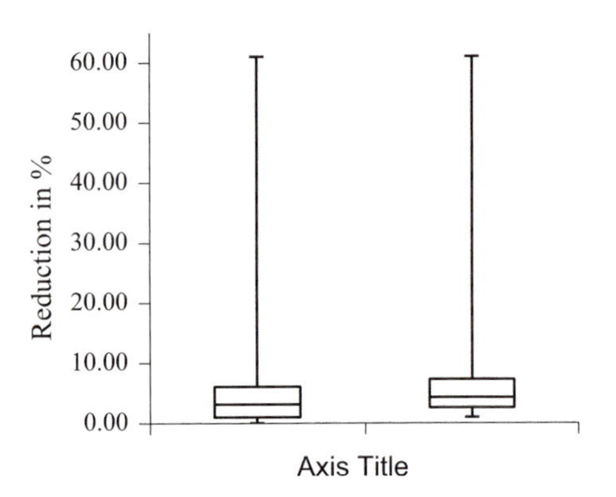

Fig. 9 Efficient route without (*left*) and with (*right*) filter

Now, only 31% of the test cases generate an efficient route which consumes less energy than the shortest. The increase of efficiency, calculated by the average of energy saving in percentage, is increased to 6% in contrast to the 2% without filter.

A detailed distribution of efficiency from the 41% more efficient routes can be seen in Fig. 8. The distribution of the energy saving in percentage is displayed in intervals. Between the parentheses, the relative frequency of the number of routes, contained in the interval, is listed. It is interesting to note that some routes save more than 50% energy. The maximum saving detected is equal to 61% with a distance of 1.8 km and a detour of 160 m.

Striking is the prevalence of the routes belonging to the first interval, which is equal to 67%. For this reason, this interval is split as shown in Fig. 11. The aforementioned phenomenon, caused by switching of the sides of the road can be observed in 9.7% of the test cases. The prevalence of this phenomenon clearly reduces the average reduction of energy consumption.

From the box plot in Fig. 9, can be observed that both quantiles are below the 10% mark. The filter cannot increase the quantiles above the 10% mark.

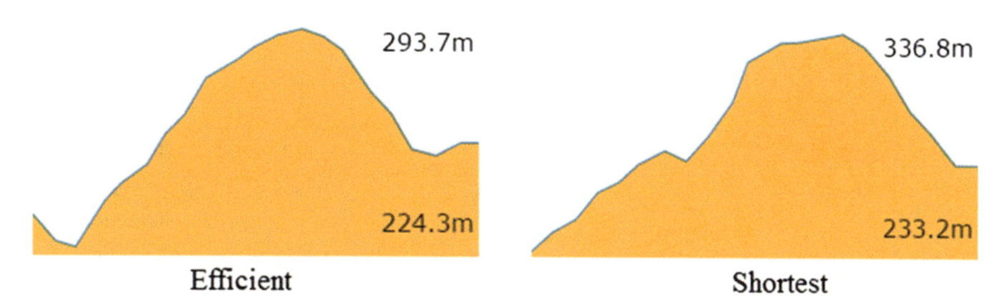

Fig. 10 Comparison of the height profile between the efficient and shortest route of an outlier

Fig. 11 Distribution of the efficient routes with up to 5% savings

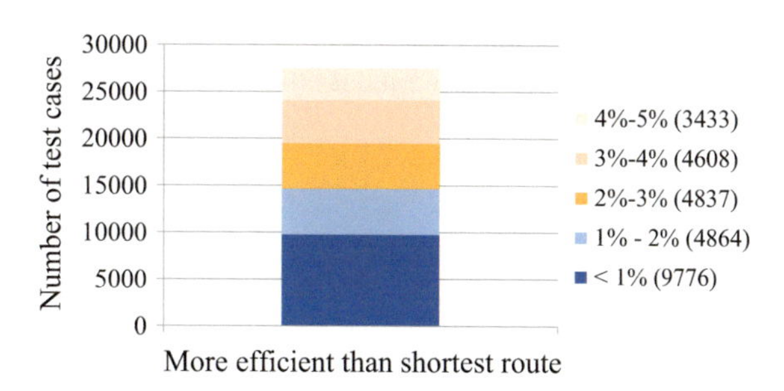

It is noteworthy that efficient route can save more than 7% for the half of the test cases, which reaffirms the benefit of the efficient routes. Especially if the short detours are taken into consideration, the savings should be evaluated positively. Noteworthy are the outliers that save more than 50% on energy consumption, Fig. 10 shows the height profile of the efficient and shortest route belonging to such an outlier. The height difference of the shortest route is clearly higher than the difference of the efficient route. The exponentiality of the consumption function (Sect. 4.2.1) explains the savings of the efficient route (Fig. 11).

The same phenomenon as with energy saving can be observed in Fig. 12, which depicts the increase in travel distance of the energy efficient route in contrast to the shortest route. Switching of the sides of the road leads to short detours of several meters and are excluded if the difference is less than 10 m.

On average, the detours required for the efficient route only increase travel distance by 1%. By applying the filter that rules out the test cases in which only the sides of the road are switched, the detours induce an increase of travel length by 5% on average. Interestingly, the average increase in travel length in percentage is very similar to the average energy saving in percentage. But no relation between the length of detour and the saving can be recognized.

The efficient route with the longest detour, increases travel distance by 180% and achieves an energy consumption reduction of 30% in contrast to the shortest route. On the other hand, an efficient route, calculated during the test, lengthens travel distance by 160%. But only 5% energy is saved in contrast to the shortest route.

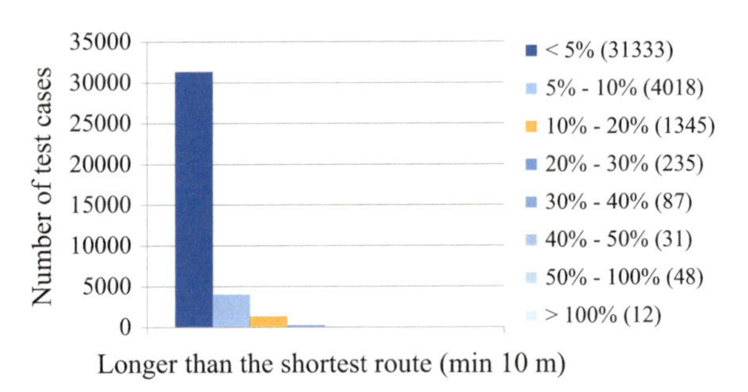

Fig. 12 Detour of the efficient route

Because of this phenomenon, it is sensible to allow the user to choose between taking the shortest or the most efficient route.

6 Conclusion

In this work, a navigation system for the disabled is proposed. The eNav project not only makes accessible routing and an increase in driving comfort for wheelchair users possible, but also enables users to get a course estimations whether the battery capacity of their electric wheelchair suffices to finish a planned trip or not.

By using a three-dimensional map, eNav can deliver the shortest and the most energy efficient route from a start to a destination point, allowing users to choose one of the two available options. During testing, efficient routes have been found which reduce energy consumption by more than 60% in contrast to the shortest route. The detour required in this case is usually less than 5% of the total length of the shortest route.

On account of the usage of crowdsourcing, the quality of accessibility information is increased permanently, utilizing manual as well as automatic data collection methods. Furthermore, the eNav App uses a surface detector, which recognizes surface types by their linear acceleration along the z-axis when a user drives over a specific type of surface. This surface information is then added to the map-data of eNav.

A next extension of eNav will be the integration of other disabled road users, like those in need of a walking frame. For this group of the disabled different requirements apply as well. The requirements can be fulfilled by using the map-data from eNav. We thank the Center for European Research on Mobility (CERM)[1] for their support.

[1]http://www.cerm.rwth-aachen.de/.

References

1. Statistisches Bundesamt: 7,5 millionen schwerbehinderte Menschen leben in Deutschland. Statistisches Bundesamt (2014). https://www.destatis.de/DE/PresseService/Presse/Pressemitteilungen/2014/07/PD14_266_227pdf.pdf?__blob=publicationFile. Accessed 26 Apr 2016
2. Franke, D., Dzafic, D., Weise, C., Kowalewski, S.: Konzept eines Mobilen OSM-Navigationssystems für Elektrofahrzeuge. In: Proc. Angewandte Geoinformatik 2011 - Beiträge zum 23. AGIT-Symposium (AGIT), Wichmann Verlag, pp. 148–157 (2011)
3. Ramm, F., Topf, J.: OpenStreetMap–Die freie Weltkarte nutzen und mitgestalten, Berlin (2010)
4. Franke, D., Dzafic, D., Weise, C., Kowalewski, S.: Entwicklung eines mobilen Navigationssystems für Elektrofahrzeuge auf basis von OpenStreetMap-Daten. In: Konferenz für Freie und Open Source Software für Geoinformationssysteme (FOSSGIS), Heidelberg (2011)
5. Franke, D., Dzafic, D., Baumeister, D., Kowalewski, S.: Energieeffizientes Routing für Elektrorollstühle. In: 13. Aachener Kolloquium Mobilität und Stadt (ACMOTE), Aachen (2012)
6. Bezirksregierung Köln (2011) Topographische Reliefinformationen, Köln
7. Norwig P., Russell, S.: Artificial Intelligence—A Modern Approach, München (2009)
8. Hart, P.E., Nilsson, N.J., Raphael, B.: A formal basis for the heuristic determination of minimum cost paths. IEEE, Trans. Syst. Sci. Cybern. Menlo Park, California, USA
9. Howe, J.: The Rise of crowdsourcin. Wired Mag. **14** (2006)
10. Kleemann, F., Voß, G.G., Rieder, K.: Crowdsourcing und der arbeitende Konsument. In: Arbeits-und Industriesoziologische Studien 1, Nr. 1, S. 29–44 (2008)
11. Gregor, K., Mlynarski, M.: Mobile Testing und Usability Testing? (2013)
12. Prandi, C., Paola, S., Silvia, M.: mPASS: integrating people sensing and crowdsourcing to map urban accessibility. In: Proceedings of IEEE International Conference on Consumer Communications and Networking Conference, pp. 10–13 (2014)
13. Menkens, C., Sussmann, J., Al-Ali, M., Breitsameter, E., Frtunik, J., Nendel, T., Schneiderbauer, T.: EasyWheel-A mobile social navigation and support system for wheelchair users. In: Information Technology: New Generations (ITNG), pp. 859–866 (2011)
14. Palazzi, C.E., Teodori, L., Roccetti, M.: Path 2.0: a participatory system for the generation of accessible routes. In: Multimedia and Expo (ICME), pp. 1707–1711 (2010)
15. Neis, P., Alexander, Z.: Openrouteservice.org is three times "open": Combining Opensource, OpenLS and OpenStreetMaps. GIS Research UK (GISRUK 08), Manchester (2008)
16. Müller, S., Kamieth, F., Braun, A., Dutz, T., Klein, P.: User requirements for navigation assistance in public transit for elderly people. In: Proceedings of the 6th International Conference on Pervasive Technologies Related to Assistive Environments, p. 55 (2013)
17. Sozialhelden, E.V.: www.wheelmap.org (2016). Accessed 26 Apr 2016
18. Dzafic, D., Klug, S., Franke, D., Kowalewski, S.: Routing über Flächen mit SpiderWebGraph. In: Proc. J. für angewandte Geoinformatik 1-2015, 516–525 (2015)
19. Dzafic, D., Franke, D., Baumeister, D., Kowalewski, S.: Modifikation des A*-Algorithmus für energieeffizientes 3D-Routing. In: Proc. Angewandte Geoinformatik 2013—Beiträge zum 25. AGIT-Symposium (AGIT), Wichmann Verlag, pp. 414–423 (2013)
20. Kothgasser, U.: Oberflächenbeurteilung von Radverkehrsanlagen mittels GPS und Beschleunigungssensoren. Universität für Bodenkultur Wien, Master thesis, Februar 2012 (2012)
21. Poppe, D.: Design and Evaluation of a Hardware Platform for Precise State-Of-Charge Determination of Lead-Acid Accumulators. RWTH Aachen University, Bachelor thesis (2011)
22. Schmitt, S., Schlender, D.: Untersuchung zum saisonalen Reifenwechsel unter Berücksichtigung technischer und klimatischer Aspekte. Bergische Universität Wuppertal, Bericht (2003)

23. Dzafic, D., Baumeister, D., Franke, D., Kowalewski, S.: Integration von Bodenbelagsinformationen zum energieeffizienten Routen von Elektrorollstühlen. In: Proc. Angewandte Geoinformatik 2014 (AGIT), vol. 26 in Beiträge zum AGIT-Symposium Salzburg, pp. 451–460 (2014)

Part IV
Technical Research for Reside and Living

Regulation of Ventilation Systems Based on Psychophysical Principles

J. Flessner and M. Frenken

Abstract This article describes an adaptive regulation method of ventilation systems based on psychophysical principles. Current standards for ventilation regulation are focused on energy consumption and comfort requirements. These ventilation methods are geared towards healthy people and leave special requirements of people with impairment out. Although recent findings suggest that people with impairment, like chronic obstructive lung disease (COPD) or asthma, are more sensitive to inadequate air quality, special requirements are not considered by current regulation standards. This paper presents a concept to improve the common regulation method which is developed to satisfy these special requirements. The concept is based on Fanger's psychophysical approach to consider the humans' perception in ventilation control. In addition to the perceived indoor air quality, the perception of one's state of health is part of the following concept. To gain knowledge about the perceived state of health, psychophysical measurements have to be carried out. The authors present a concept for an adaptive ventilation control, which is able to take special requirements of elderly and people with impairment into account.

1 Introduction

High concentrations of indoor air pollutants can lead to lower perceived comfort and unhealthy environmental conditions. The ventilation of the indoor environment is substituting the polluted indoor air with less polluted outdoor air. Polluted air can have different negative effects on the inhabitant, like sensory irritation or respiratory illness [14]. Therefore, the exposure period of the inhabitant to unhealthy air quality should be minimized. Adequate air change rates are important especially as humans spend nearly 90% of the day inside buildings in average [3]. Thus, the humans are likely to be exposed to the environmental factors for the most part of the

J. Flessner (✉) · M. Frenken
Institute of Technical Assistance Systems, Jade University of Applied Sciences, Ofener Street 16/19, 26121 Oldenburg, Germany
e-mail: jannik.flessner@jade-hs.de

R. Wichert and B. Mand (eds.), *Ambient Assisted Living*, Advanced Technologies and Societal Change,
DOI 10.1007/978-3-319-52322-4_10

day. According to this, it is possible to influence the humans' condition and state of health through an intelligent regulation of ventilation.

The standards for ventilation of non-industrial buildings are designed to be energy efficient and use static thresholds to establish an adequate level of comfort. Nevertheless, the air change rate per person differ from 4 l/s per person to 25 l/s per person in the different European standards [3]. Therefore, a uniform regulation method based on the humans' requirements could lead to an adapted and healthier building environment. In order to identify the health effects of different air change rates, a group of scientists reviewed several studies in the early '00s [13]. As a result, they associated the ventilation strongly with the symptoms of sick building syndrome (SBS), inflammation, infections, asthma, allergy, short-term sick leave and the perception of air quality.

The awareness of the utility of psychophysics for system design raised in recent years. Especially the psychophysical measurements of tactile stimuli were used for the design of human machine interaction. In this context, the adequate frequency and position of vibro-tactile displays or necessary temperatures for thermo-tactile displays were examined with psychophysical principles [7, 11]. These examples show, that the human machine interaction could benefit from the involvement of the humans' perception through psychophysical principles. According to this, the home automation, including the ventilation control, could benefit from integrating psychophysical measurements.

Therefore, a human-centered design of the ventilation control, involving the individual sensitivity on indoor air quality, may be a way to improve the building environment. Through the involvement of the inhabitants' condition in the ventilation control process, an improved human building interaction could be established. This process is pictured in Fig. 1 and illustrates the interaction between the inhabitant and the ventilation control. In present days the ventilation control uses mainly static threshold values from given guidelines and do not react on changes in the building environment. In case the inhabitants want to manipulate the performance of the ventilation system to satisfy their requirements, they have to interact directly through an input option. Otherwise the ventilation control system does not react on the actual condition and requirements of the inhabitants. Therefore the addition of a control feature, which reacts to the inhabitant's condition, may enhance the interaction between the inhabitant and the ventilation control.

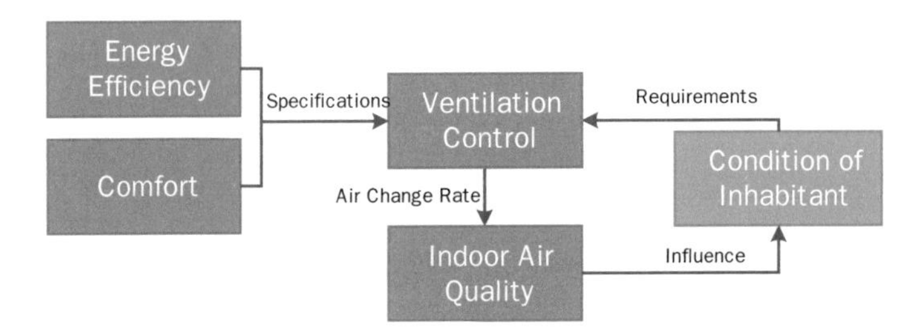

Fig. 1 The current ventilation control process and the consideration of the human

2 Related Work on Ventilation Control and Health

A method to regulate the indoor air quality through the measurement of the CO_2 concentration was already established in the year 1858 by Pettenkofer [1]. This approach is based on the assumption, that the human is the only pollution source in non-industrial buildings. Hence, the air change rate is dependent on the number of inhabitants. However, studies concerning the pollution sources proved that the indoor air quality is affected by far more environmental objects. In order to evaluate the response of people on the air quality affected by different sources, several studies were performed. These psychophysical experiments contained the rating of the air quality with different scales and a large number of test persons [12, 15].

Fanger used the results of his studies to establish two measurement units for the perception of air quality: olf and dezipol [6]. The Olf describes the strength of a pollution source. A sitting average adult worker in an office is rated with 1 olf. The unit dezipol is used to measure the perceived air quality. One dezipol is reached by a room contamination of 1 olf and an air change rate of $10\,\mathrm{l/s}$. In conclusion, the concept of Fanger combines the measurement of the air quality perception, which are influenced by psychological and physical parameters, and the regulation of the indoor air quality. Fanger's concept is used as basic principle in common standards, e.g. DIN EN 15251 [5], to regulate the air change rate based on the room dimensions or number of inhabitants.

The perception of the indoor air quality is dependent on the psychological and physical condition of the inhabitant. Wargocki pointed out that the rating of the indoor air quality is dependent on different sensory stimuli [12]. Moreover, Wargocki concluded that the evaluation of indoor air quality is related to individual parameters.

Recent studies showed that the sensitivity of elderly and people with respiratory impairment to air pollutants is heightened. Hansel and colleagues studied the effects of different air pollutants on patients with COPD [8]. It was shown that even low concentrations of air pollutants like particulate matter and nitrogen dioxide are associated with increased symptom appearance. These findings imply that patients with COPD have special requirements on the indoor air quality.

Furthermore, Bentayeb et al. researched the relation between the respiratory health of elderly and indoor air quality [2]. They associated the appearance of respiratory illness symptoms with inadequate indoor air quality as a result.

Wargocki et al. [13] reviewed findings about the effect of ventilation on health. As a result the scientists recommended an air change rate above $25\,\mathrm{l/s}$ to provide a healthy indoor air quality. Furthermore, Sundell and colleagues confirmed the statement of positive health effects through an air change rate above $25\,\mathrm{l/s}$ [10]. These reviews do not differentiate between the individual state of health and attempted to give general recommendations. The effect of impairments on the requirements on indoor air quality was not surveyed.

These findings show the potential of an adaption of the ventilation control concerning special requirements. Nevertheless, there is a lack of knowledge about the perception of indoor air quality by people with impairment. Fanger's approach to measuring the perception of the air quality involved a large group of young and healthy people [6]. The psychological or physical condition of the study subjects was not distinguished. The method is a general approach for a human centered ventilation design. Therefore, an extension of Fanger's air change rate calculation concerning the special requirements of elderly and people with impairment may lead to an adaptive and individualized method to regulate the indoor air quality.

3 Adaptive Ventilation Control

Based on the motivation to design a individualized and adaptive method for indoor air quality control, a novel concept will be described in the following chapter. According to this, the approach of Fanger [6] is the basic principle of the adaptive method. To clarify Fanger's method for the estimation of an adequate air change rate, an example for an office room is presented. Furthermore, an approach to extent this method for people with special requirements on indoor air quality is outlined.

3.1 Fanger's Concept for Air Change Rate Calculation

To calculate the required air change rate V with Fanger's method (Eq. 1), the outdoor air quality c_{out}, the desired indoor air quality level c_r and the total air pollution G_i has to be defined.

$$V = 10\frac{\sum_i G_i}{(c_r - c_{out})} \tag{1}$$

The total air pollution G_i can be calculated through the number of inhabitants and the environmental materials. Because of missing information about the pollution caused by several materials, the pollution though environmental materials can be estimated by the room type and dimensions. An overview of the different parameters for the estimation of G_i is given in Table 1. These parameters contain information about the inhabitants' smoking behavior and the usage of the room. The definition of the room types leads to an expected mean room occupancy and mean pollution according the room dimensions [9].

Following example will demonstrate the application of Fanger's method to calculate the required air change rate for a 20 m^2 non-smoker office:

- Assumptions for G_i: Smoking and activity factor equals 1 $^{olf}/_{person}$, the mean room occupancy equals 0.07 $^{olf}/_{m^2}$, the mean room pollution equals 0.3 $^{olf}/_{m^2}$.
- Assumption for c_r: A high air quality is desired ($c_r = 0.6$ dezipol)
- Assumption for c_{out}: The outdoor air quality is high ($c_{out} = 0.1$ dezipol)

According to Fanger's method, an air change rate of 148 $^{m^3}/_{h}$ is required to provide a high air quality for a 20 m^2 non-smoker office. On the assumption, that two

Table 1 Parameter for the calculation of the total air pollution G_i [9]

Smoking factor [olf/person]		Activity factor [olf/person]		Mean room type occupancy [persons/m²]		Mean room type pollution [olf/m²]	
0 cigarettes/h	1	Sitting	1	Office	0.07	Office	0.3
0.24 cigarettes/h	2	Low	4	Meeting room	1.5	Meeting room	0.5
0.48 cigarettes/h	3	Mid	10	Classroom	0.5	Classroom	0.3
1.2 cigarettes/h	6	High	20	Kindergarten	0.5	Kindergarten	0.3
				Living room	0.05	Desired	0.1

people are working in this office, a comparison with the current European standard for non-residential buildings EN 13779 leads with 144 $^{m^3}/_{h}$ to a nearly equal result [4]. The calculation was performed for the best air quality category described in the standard EN 13779. According to this category, an air change rate of 20 $^{l}/_{s}$ pro person is required to obtain the highest indoor air quality. Hence, the standard for non-residential buildings and the method concerning the humans' perception show insignificant differences. So, people without the need of an air change rate above the current standards and with higher priority on energy efficiency are as well represented in Fanger's method. Therefore, the approach to regulate the ventilation based on psychophysical principles can be seen as a legitimate basis to develop a more adaptive regulation method.

A novel regulation method is motivated through the fact that special requirements caused from impairments and health issues are not considered, neither in current standards nor in Fanger's approach. The main purposes of these methods are to secure comfort and to create an energy-saving ventilation. Whereas, Fanger's approach is focused on the involvement of the indoor air quality perception.

3.2 *Extension of Fanger's Concept Concerning Special Requirements*

The comparison between the recommended air change rate above 25 $^{l}/_{s}$ and the air change rates calculated by standard EN 13779 and by Fanger's approach, indicate the potential to create healthier environments. A regulation based on common standards may be sufficient for people without impairment to satisfy their comfort requirements. Nevertheless, for people with impairments the state of health is of higher priority than energy efficiency issues. In order to involve people with special requirements a novel approach has to be followed. The concept of this approach will be described in the following. The central question is, how knowledge about people with special requirements on the indoor air quality can be acquired and respected in an extension of Fanger's method.

The basic information of Fanger's approach for the calculation of adequate air change rates were received from psychophysical measurements of the perceived indoor air quality. The results were used to establish a link between the perceived indoor air quality and air change rate. Nevertheless, beside the activity factor and the smoking behavior, Fanger's method does not involve the physical and psychical condition of the inhabitants. A relation between the state of health and the air change rate were not introduced. To extend Fanger's approach with a differentiation between healthy people and people with impairment, this relation had to be established.

To develop a method for ventilation control, which reacts to the inhabitants' state of health, psychophysical measurements are necessary to link the perceived state of health with air change rates. As shown before (Sect. 2), the appearance of illness symptoms is related to inadequate indoor air quality. According to this, the perceived

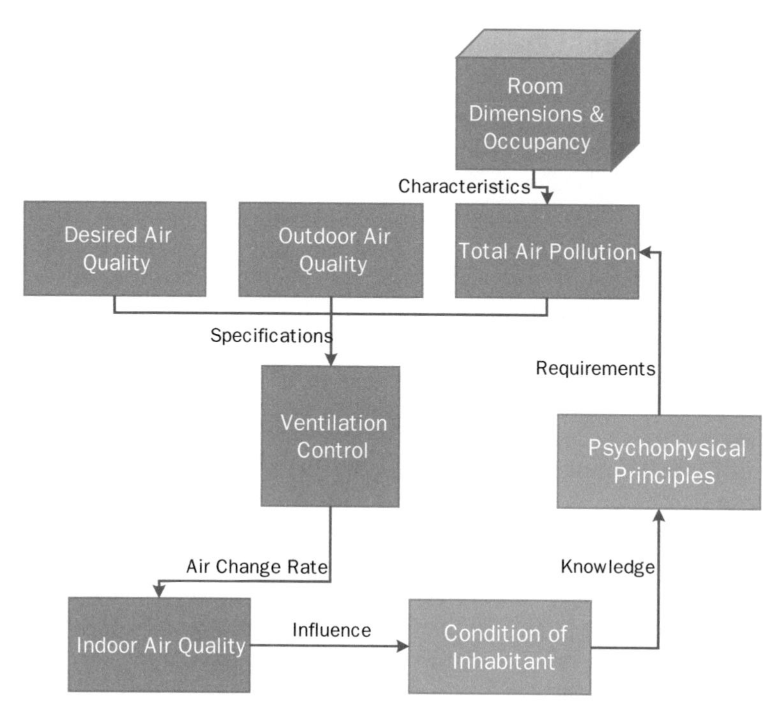

Fig. 2 The ventilation control process according to Fanger and its' human centered extension

state of health can be linked to the regulation of the air change rate. Therefore, groups of people with higher sensitivity on inadequate indoor air quality have to be identified and the perceived state of health in environments provided with different air change rates have to be evaluated.

The knowledge received from these psychophysical measurements concerning the state of health can be used to extent Fanger's approach. In order to achieve this a parameter representing the state of health can be introduced. One way to achieve an individualization of the air change rate is the addition of a novel parameter to the total air pollution G_i of Fanger's method, similar to the smoking behavior (Fig. 2). A new parameter should, according to the results of these studies, involve the type and the intensity of the impairment. The intensity of impairments can be of importance to not overstate the air change rate with respect to the energy efficiency.

Another option is to adapt the desired air quality value c_r (Eq. 1) with respect to the study results. The desired air quality is linked to the percentage of expected unsatisfied. Because of the probably increased appearance of illness symptoms, the perception of the own state of health is linked to the current air quality. It can be assumed that people with increased sensitivity to indoor air pollutants are more likely to be dissatisfied with low air change rates. Therefore, the adaption of the desired indoor air quality value is able to represent the inhabitants' needs. The increase of the value based on the type and intensity of the respective impairment could be a solution for the adaption of the ventilation control.

4 Conclusion and Outlook

Current standards in ventilation control left the requirements on indoor air quality of people with impairment out. Although scientific works recommend an air change rate of 25 l/s to create healthy environments, applicable standards focus the control process on energy efficiency and comfort. Recent studies showed the increased sensitivity of people with respiratory impairment on even low concentrations of indoor air pollutants. Hence, these people have special requirements on indoor air quality. The authors were able to present a novel concept that takes account of these studies. This concept identifies adequate air change rates based on psychophysical information. The health aspect should have the highest priority in the design of ventilation regulation.

The next steps of the authors will be to include the special requirements of people with impairment by measuring the relation between the state of health perception and air change rates. Thus, a human centered approach established by Fanger et al. [6] was chosen as basis process. The extension of this human centered approach with the inclusion of psychophysical principles respecting the perceived state of health may afterwards lead to an improved interaction between the inhabitant and his environment. Furthermore, this concept can possibly be used to improve the parametrization of entire home automation. The involvement of psychophysical measurements and knowledge in home automation control opens up new possibilities to create improved and healthy environments.

References

1. von Pettenkofer, M.: Über den Luftwechsel in Wohngebäuden. Cotta (1858)
2. Bentayeb, M., Norback, D., Bednarek, M., Bernard, A., Cai, G., Cerrai, S., Eleftheriou, K.K., Gratziou, C., Holst, G.J., Lavaud, F., et al.: Indoor air quality, ventilation and respiratory health in elderly residents living in nursing homes in Europe. Eur. Respir. J. ERJ, 00,824 (2015)
3. Brelih, N., Seppänen, O.: Ventilation rates and IAQ in european standards and national regulations. In: The Proceedings of the 32nd AIVC Conference and 1st TightVent Conference in Brussels, pp. 12–13 (2011)
4. Din, E.: 13779: Lüftung von nichtwohngebäuden-allgemeine grundlagen und anforderungen an lüftungs-und klimaanlagen. Beuth, Berlin (2005)
5. Din, E.: 15251: 2007–08: Eingangsparameter für das raumklima zur auslegung und bewertung der energieeffizienz von gebäuden-raumluftqualität. Temperatur, Licht und Akustik (2007)
6. Fanger, P.O.: Introduction of the olf and the decipol units to quantify air pollution perceived by humans indoors and outdoors. Energy Build. **12**(1), 1–6 (1988)
7. Gallo, S., Cucu, L., Thevenaz, N., Sengul, A., Bleuler, H.: Design and control of a novel thermo-tactile multimodal display. In: 2014 IEEE Haptics Symposium (HAPTICS), pp. 75–81. IEEE (2014)
8. Hansel, N.N., McCormack, M.C., Belli, A.J., Matsui, E.C., Peng, R.D., Aloe, C., Paulin, L., Williams, D.L., Diette, G.B., Breysse, P.N.: In-home air pollution is linked to respiratory morbidity in former smokers with chronic obstructive pulmonary disease. Am. J. Respir. Crit. Care Med. **187**(10), 1085–1090 (2013)
9. Rietschel, H., Esdorn, H.: Raumklimatechnik: Grundlagen. Springer (2008)

10. Sundell, J., Levin, H., Nazaroff, W.W., Cain, W.S., Fisk, W.J., Grimsrud, D.T., Gyntelberg, F., Li, Y., Persily, A., Pickering, A., et al.: Ventilation rates and health: multidisciplinary review of the scientific literature. Indoor Air **21**(3), 191–204 (2011)
11. Van Erp, J.B.: Guidelines for the use of vibro-tactile displays in human computer interaction. In: Proceedings of Eurohaptics, vol. 2002, pp. 18–22 (2002)
12. Wargocki, P.: Measurements of the effects of air quality on sensory perception. Chem. Sens. **26**(3), 345–348 (2001)
13. Wargocki, P., Sundell, J., Bischof, W., Brundrett, G., Fanger, P.O., Gyntelberg, F., Hanssen, S., Harrison, P., Pickering, A., Seppänen, O., et al.: Ventilation and health in non-industrial indoor environments: report from a european multidisciplinary scientific consensus meeting (euroven). Indoor Air **12**(2), 113–128 (2002)
14. World Health Organization: WHO Guidelines for Indoor Air Quality: Selected Pollutants. WHO (2010)
15. Zhang, H., Arens, E., Kim, D., Buchberger, E., Bauman, F., Huizenga, C.: Comfort, perceived air quality, and work performance in a low-power task-ambient conditioning system. Build. Environ. **45**(1), 29–39 (2010)

Computer-Based Adaption of Cooking Recipes Integrated in a Speech Dialogue Assistance System

Karen Insa Wolf, Stefan Goetze and Frank Wallhoff

Abstract Speech input and output allow for a natural and intuitive means to communicate with a technical device or system. This motivates the development of speech dialogue systems for convenient human-machine interaction. A speech dialogue system is particularly suitable in scenarios in which the user cannot use his or her hands for interaction, for example while driving a car. In kitchen environments, a speech dialogue system promises to be very useful as well. The application CooCo (Cooking Coach) aims at providing assistance under working conditions in a common kitchen. CooCo supports the user to choose recipes best matching the requests of the user and gives advice during the cooking processes. In this paper, the focus lies on the conceptual view on building up CooCo and the integration of a computer-based approach to adapt cooking recipes.

1 Motivation

Nowadays, speech dialogue systems have become more and more common in everyday use, even for people with a low affinity to new technologies. A speech dialogue system is particularly suitable in scenarios in which the user cannot use his or her hands for interaction, for example while driving a car (cf. e.g. Larsson and Villing [11]). Speech dialogue systems promise to be very useful as well during daily work

K.I. Wolf (✉) · S. Goetze (✉) · F. Wallhoff (✉)
Fraunhofer Institute for Digital Media Technology IDMT, Marie-Curie-Strasse 2,
26129 Oldenburg, Germany
e-mail: insa.wolf@idmt.fraunhofer.de

S. Goetze
e-mail: stefan.goetze@idmt.fraunhofer.de

F. Wallhoff
e-mail: frank.wallhoff@idmt.fraunhofer.de; frank.wallhoff@jade-hs.de

F. Wallhoff
Jade University—Institute for Technical Assistive Systems, Westerstr. 10-12,
26121 Oldenburg, Germany

R. Wichert and B. Mand (eds.), *Ambient Assisted Living*,
Advanced Technologies and Societal Change,
DOI 10.1007/978-3-319-52322-4_11

at home in the kitchen. The user can ask for recipes while doing the dishes or can get reminders regarding timing and next steps from an assistance system while cooking. Assuming a flexible dialogue management, spontaneous utterance of the user (like e.g. ≪Oops, I do not have ...≫) can be processed.

The nutrition topic itself is of particular relevance in context of the demographic change and, thus, for ambient assisted living. Well-balanced nutrition is an important pillar of a healthy lifestyle and healthy ageing (cf. e.g. Zbeida et al. [21]). Nutrition consulting can contribute to the prevention and therapy of diseases thereby enabling a longer, independent life. With increasing age, the need for a supporting nutrition consultation can grow, as chronic diseases become more frequent. Depending on the health status of a person, cognitive impairments such as memory decline or decline in the capacity to structure ones daily routine increase, both of which affect the ability to independently schedule food intake. Loss of the partner can also induce a less balanced nutrition.

Contributing to healthy ageing and long-lasting independence of older people additionally motivates the development of the speech dialogue system CooCo (Cooking Coach), introduced in Wolf et al. [20]. In this paper, the focus lies on the conceptual view on building up CooCo and the integration of a computer-based approach to adapt cooking recipes. The automatic adaption of recipes can be helpful to enlarge the recipe database for special diet purposes (vegetarians, allergic persons).

2 Concept of CooCo

The application CooCo aims at providing ambient and non-dominant meal preparation assistance under real-world conditions. Keyboard, mouse or touchscreens are no practical user interfaces while cooking. Therefore, CooCo uses automatic speech recognition and speech output as the user interface. Two different use cases are considered in the design of the system.

First, in the use case *recipe advice*, CooCo helps the user to choose the right recipes based on her or his requests concerning ingredients, preparation time, complexity of preparation or even tastes (e.g. sweet or hot). Over time CooCo learns the likes and dislikes of the user to optimize the interaction. The recipe advice mode includes generic models of gustatory preferences, e.g. hot or sweet depending on typical amount of ingredients like chili or sugar, and a computer-based adaptation of the cooking recipes, cf. Wolf et al. [19].

Second, in the use case *cooking support*, CooCo supports the user during the cooking process. The goal thereby is not only to read the cooking steps as shown e.g. in Schäfer et al. [16], but to build up a plan considering expenditure of time and dependencies of single cooking steps. CooCo formulates an action plan considering active time of the user, like cutting, and passive time, like simmering.

Both tasks, recipe advice and cooking support, require a context-based dialogue system including modules for interpreting, planning and re-planning, memorizing and learning. These modules are typical for a cognitive (technical) system. Such systems can represent system-relevant aspects of the environment internally, which is the basis for cognitive functions, cf. Gamrad [6], and allows flexible behaviors.

In this paper, just the first use case is addressed in more detail to explain the integration of the new feature of automatic adaption of cooking recipes. The concept of the second use case is explained in more detail in [20].

3 Modules of CooCo

The system of CooCo is based on seven main modules, as explained in the following. A schematic diagram of the architecture of CooCo is shown in Fig. 1. *Speech recognition* and *text to speech* (TTS) modules are introduced as part of a dialogue system in Sect. 3.1. Details for the *dialogue manager* are given in Sect. 3.2. The *database* is introduced in Sect. 3.3. The approach of adaption of cooking recipes as part of the *learning unit:facts* is explained in Sect. 4. The *dynamic process manager* is conceptually based on a scheduler, handling time processes, and a process planner which derives a plan for the cooking process regarding dependencies. The concept of the dynamic process manager is explained in more detail in [20]. In future versions, the *learning unit:processes* should adapt parameters of the cooking action description to match better the preferences and practices of the user.

3.1 Dialogue Systems

In Fig. 2 the architecture of a speech dialogue system is sketched, cf. e.g. Eliasson [5]. The speech recognizer translates from spoken input to written text. The interpretation of the utterance is encapsulated in the module *dialogue manager* based on written

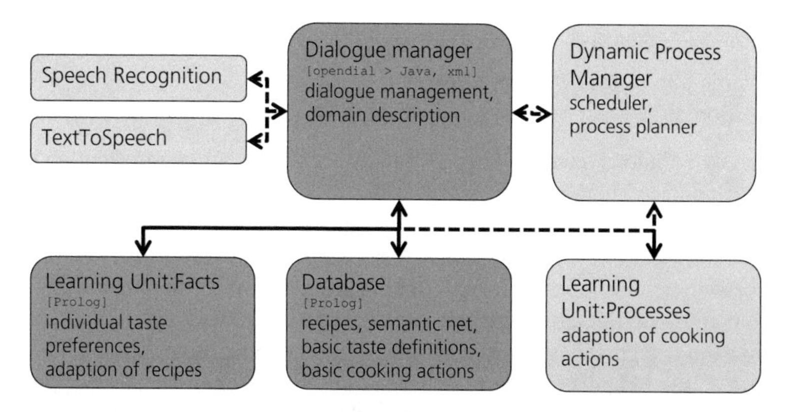

Fig. 1 A schematic diagram of the architecture of CooCo. The modules in dark gray are active in the application discussed in this paper

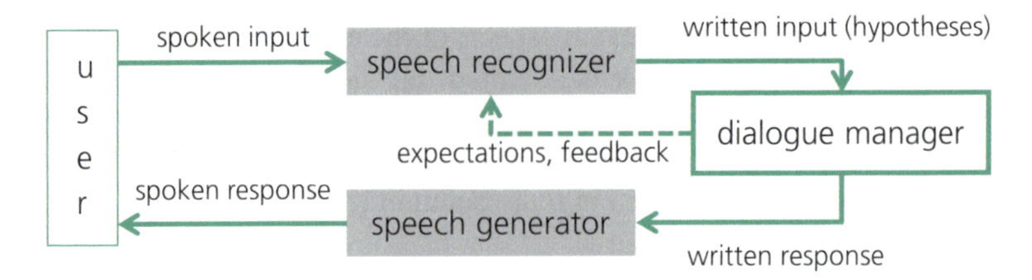

Fig. 2 Architecture of a speech dialogue system with an optional feedback between dialogue manager and speech recognizer

text input. To improve speech recognition results, expectations of the content of the user utterances can be introduced in the speech recognizer. A speech generator (text-to-speech TTS) is used to articulate the response.

Focused on dialogue management, speech recognition is often linked as an off-the-shelf module in the system, cf. e.g. Eliasson [5]. Different systems are available for this purpose, some of them are able to return not just one best hypothesis but a ranked n-best list of hypotheses for further context based analysis, cf. Morbini et al. [13]. For CooCo automatic speech recognition (ASR) is used, cf. Goetze et al. [7]. Therefore, specific signal enhancement techniques are necessary, cf. e.g. Rennies et al. [15]. A feedback loop between the dialogue manager and the speech recognizer is designed in the concept based on different input hypotheses entered in the dialogue manager to improve speech recognition. The feedback information of the chosen most likely recognition results and expectations are used to weight the probability of the words in the dictionary of the recognizer. This helps optimizing the recognition rate.

3.2 *Dialogue Manager*

The tasks of the dialogue manager can be summarized as

1. interpretation of the (written) utterance
2. definition of next action resulting in a response (question, information, advice)

These tasks include diverse subtasks as parsing, updating actual information knowledge, or planning. There are different approaches to solve these tasks depending on the specific use case and necessary functionality of the dialogue manager, cf. e.g. Morbini et al. [13]. For CooCo's first task, giving recipe advice, a slot-filling system would be sufficient. The second task, cooking support, is more demanding if the aim is more than just articulating the cooking steps when the user asks for a next step. Applying the categories of Morbini et al. [13], a negotiation and planning system is needed. Lison [12] distinguishes between hand-crafted and statistical approaches for a speech dialogue manager and proposes the toolkit OpenDial [14] to combine both. Pure statistical approaches as partially observable Markov decision processes

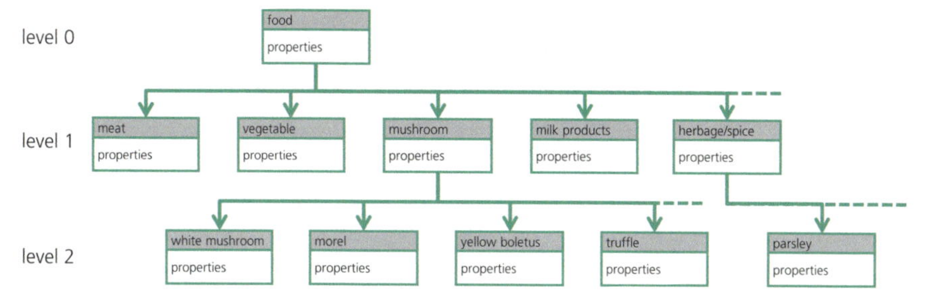

Fig. 3 Ingredients structured in a semantic net

(POMDPs), described in Williams [18], need training data sets to build up the dialogue setting. Central element of OpenDial is an information-state which is updated cyclically based on the information of the user (and maybe from other sources, e.g. sensors). The dialogue manager of CooCo is based on OpenDial. Another option for an information-state approach is the TRINDI Concept (task oriented instructional dialogue, Larsson and Traum [10]), as proposed in Wolf et al. [20], but the basic algorithm does not consider probability and utility values to influence the dialogue policy. Therefore, the approach of Lison [12] is used.

3.3 Database

In the knowledge database facts are stored: recipes, ingredients, basic taste definitions and basic cooking actions. The *database* currently contains 1.222 recipes [8]. The ingredients are stored as a semantic net as sketched in Fig. 3. This network is used for different tasks. Assuming the user requests a recipe with meat, CooCo can resolve this request by looking for all recipes using chicken, turkey, etc. as ingredients. A shopping list can be generated in a sorted form for a well-organized shopping tour, e.g. all vegetables are written in one block of the list. Finally, properties based on the inheritance relations in the network are applied within the algorithm to adapt the cooking recipe, as described in Sect. 4.

4 Computer-Based Adaption of Cooking Recipes

The computer-based variations of cooking recipes addresses topics of artificial intelligence and machine learning approaches, cf. e.g. Cordier et al. [1]. The task to derive the consequences of the substitution of an ingredient on the textual description of the preparation steps requires techniques of natural language understanding, e.g. Dufour-Lussier et al. [3]. Other approaches aim at replacing ingredients, e.g. by randomizing recipe items [4], by using cognitive super computing (based on IBM's computer system WATSON, [9]) or by just enlarging the database (by the help of a community) to find a matching recipe for every combination of ingredients [17].

4.1 Basic Approach

In Wolf et al. [19] a heuristic computer-based approach is described to vary cooking recipes by replacing ingredients. The approach is based on a scoring system. The score value is used to rate different ingredients as candidates to substitute a specific ingredient of a recipe. This substitution score depends on different factors: (1) rating of the similarity between the ingredient which has to be replaced and the substitution candidate, (2) rating how well the substitution candidate fits the recipe, and (3) gustatory preferences of the user. The substitution candidate with the highest score is proposed to the user.

The rating of the similarity between the ingredients is based on the one hand on their level and their distance as elements in the semantic net, shown in Fig. 3 and on the other hand on a comparison of their nutrition values. Confer dAmato et al. [2] for a detailed definition of similarity measures in the context of ontologies. The similarity values, defined here, depend on the structure of the semantic net and the specified nutrition values. How well a substitution candidate fits the recipe is derived from statistical numbers based on the recipe database. Finally, users differ in their gustatory preferences, one likes more traditional recipes, while the other is more open to new tastes. Therefore, the gustatory preferences of the user are considered by two user parameters, referred to as experimental levels: The experimental level e_{cd} influences how common or uncommon a substitution candidate should be. The level e_{cb} regulates how common or uncommon the combination of a substitution candidate and all remaining elements of the chosen recipe is. For both levels three adjustment steps can be chosen by the user, ranging from 1 = very common to 3 = very uncommon.

4.2 Exemplary Results

The approach described above is conceptually tested, further implementation and evaluation is ongoing work. The starting point is a simple mushroom soup recipe:

250 g	commonmushrooms,	40 g	butter,
40 g	flour,	5 dl	bouillon,
5 dl	milk,	– –	salt,
1 tb	parsley, minced,	– –	pepper

Numerical results for the different experimental levels e_{cd} and e_{cb} are listed in Table 1. The results show, that a user with a low e_{cd} of 1 and a medium or high e_{cb} of 2 or 3 will be recommended a tomato soup. In case a very common combination of

Table 1 Numerical results[a] of adapting a cooking recipe. The result $s(j,k)$ means s based on $e_{cd} = j$, $e_{cb} = k$. The respective candidate with the largest score s is marked in bold letters. Only some examples of the possible substitution candidates are listed

	Y. boletus	Morel	Truffle	Red pepper	Tomato	Cucumber	Cauliflower
$s(1,1)$	46	**56**	14	44	51	8	43
$s(1,2)$	46	36	34	54	**61**	28	43
$s(1,3)$	46	16	54	64	**71**	48	43
$s(2,1)$	56	**76**	34	24	31	18	53
$s(2,2)$	56	**56**	54	34	41	38	53
$s(2,3)$	56	36	**74**	44	51	58	53
$s(3,1)$	66	**96**	54	4	11	28	63
$s(3,2)$	66	**76**	74	14	21	48	63
$s(3,3)$	66	56	**94**	24	31	68	63

[a]The numbers differ slightly to those published in Wolf et al. [19] due to modifications of some of the nutrition values of the ingredients

ingredients is wanted ($s(1,1)$), morel soup is proposed instead. The reason for this is that the recipes with common mushrooms and morel often share the basic combination of ingredients. Obviously, the results depends strongly on the content of the database. A user who wants uncommon ingredients in an uncommon combination gets truffle as substitution candidate ($s(3,3)$). Elements of the class mushrooms are mostly preferred. A whole class like "mushrooms" could also be excluded, resulting in recommendations of cauliflower as substitution candidate as a less common ingredient than tomatoes. If a high experimental level e_{cb} for the final combination in the recipe is chosen you get even cucumber as candidate.

4.3 *Use Cases*

The central task in the following two use cases is to propose a tasty recipe based on the user's input by replacing ingredients. The intention of the user differs in the scenarios. Both use cases can be extended by including the question of undesired ingredients. In order to enlarge the number of possible recipe candidates, the proposed recipe variation approach can be applied in this case additionally to substitute undesired ingredients.

4.3.1 Use Case 1: ≪Surprise Me.≫

Based on one chosen recipe the user asks for a variation of this recipe. A similar scenario is that the user realizes that one of the ingredients is missing but s/he still wants to cook the chosen recipe accepting the variations. In both cases, CooCo can choose freely possible substitution candidates. In the first case, the ingredient to be substituted is not defined by the user. In the second case, this ingredient is the missing one.

4.3.2 Use Case 2: ≪Work with What I Have.≫

The user specifies some ingredients, s/he wants to work with, but no recipe can be found in the database which uses all desired ingredients.[1] The task for CooCo is now to propose one recipe which matches by replacing missing ingredients of the recipe with those defined by the user. For this scenario, a plausibility check is performed since not each combination of ingredients presents a suitable option for a recipe.

5 Implications for the Speech Dialogue Plot

Both use cases, described in Sect. 4.3, demand different plots of the speech dialogue. The first use case (≪Surprise me.≫, cf. Sect. 4.3.1) can be integrated into a simple version of a dialogue by adding the question whether a recipe should be adapted if wanted or necessary. As CooCo is more or less free to choose the candidate which substitutes one of the ingredients of the recipe, the chance is large that an appropriate solution will be found. It is different in the second use case (≪Work with what I have.≫, cf. Sect. 4.3.2) as there, the set of ingredients which can be used as the candidate for the substitution is very limited. CooCo has to discuss with the user what would be the best option starting from probably not ideal substitution solutions. An example:

In use case 2 the user desires a recipe with the ingredient set I_{us} = {*butter, flour, parsley, bouillon, redpepper*}. The recipe that matches I_{us} best is mushroom soup, given in Sect. 4.2, based on the simple rule to look for those recipes with the smallest number of missing ingredients I_{ms}. However, red pepper is not part of the original recipe. Based on the substitution score it is then checked whether red pepper is a suitable substitution candidate c_{sb} for one of the missing ingredients.

Aiming at an efficient dialogue an assumption is introduced that some ingredients are standard ingredients, e.g. {*pepper, salt, bouillon*}. They are supposed to be available also in case the user did not mention them explicitly. But this first guess has to be confirmed by the user. Then, the only missing ingredients left are I_{ms} = {*common mushrooms, milk*}. Considering the experimental levels, the score s is derived for all pairs of c_{sb} with one of the elements of I_{ms}. The highest score $s = 54$ is reached for the experimental level $e_{cd} = 1$. Considering a threshold scheme of $[120 \ldots 80]$ (very good), $[80 \ldots 40]$ (acceptable), $[40 \ldots 0]$ (not recommended) for s, the substitution

[1]The current version does not consider available amounts of items, but this will be included in future versions.

pair *red pepper—common mushrooms* is evaluated as "acceptable". In no case it is an option to replace *milk* with *red pepper*, the highest score is $s = 29$. *Milk* remains here as missing candidate. Two different last options are possible:

(1) Ask the user explicitly whether there is after all a potential substitution candidate. If yes, repeat the procedure.
(2) Evaluate how well the missing ingredient could be omitted. For this, check if other ingredients could make up for the omission by increasing its quantity based on their similarity. In this specific example, the result of 17.5 for *milk* in relation to *butter* is not promising enough to propose this as solution. As final step, the amount of liquid within the recipe ingredients is checked leading here to an increase of the amount of bouillon to recover the original amount of liquid.

The final solution with appropriate comments based on the score s is presented to the user. The example shows that there are some intermediate dialogue steps necessary to get the final result of the adapted recipe. Therefore, current work considers user study to support the definition of appropriate user dialogue scripts.

6 Discussion and Conclusion

This paper describes the concept of the speech dialogue assistance system CooCo that aims to support users during daily work at home in the kitchen. An approach to derive recipe variations by replacing ingredients is introduced. Two different use cases are addressed. The presented examples provide reasonable substitution results. However, the test cases have to be enlarged in further evaluation steps. The adaption algorithm is an initial version to include such a feature within the speech dialogue to model a more complex interaction with the user. The algorithm is still limited and may propose uncommon or nonsense substitution suggestions. But its implementation is done easily and with less effort than manually listed potential substitution pairs. The feature provides a good experimental environment to design and test speech dialogue approaches for practical use even based on the simple use case of *recipe advice*.

References

1. Cordier, A., Dufour-Lussier, V., Lieber, J., Nauer, E., Badra, F., Cojan, J., Gaillard, E., Infante-Blanco, L., Molli, P., Amedeo, N., Skaf-Molli, H.: Taaable: a case-based system for personalized cooking. In: Montani, S., Jain, L.C. (eds.) Successful Case-Based Reasoning Applications-2, Studies in Computational Intelligence, vol. 494, pp. 121–162. Springer (2014)
2. dAmato, C., Staab, S., Fanizzi, N.: On the influence of description logics ontologies on conceptual similarity. In: Knowledge Engineering: Practice and Patterns, pp. 48–63. Springer (2008)
3. Dufour-Lussier, V., Le Ber, F., Lieber, J., Nauer, E.: Automatic case acquisition from texts for process-oriented case-based reasoning. Inf. Syst. **40**, 153–167 (2014)

4. Easierbaking: The Clickable Recipe Maker (2015). http://www.easierbaking.com. Accessed 9 July 2015
5. Eliasson, K.: The use of case-based reasoning in a human-robot dialog system. PhD thesis, Linköping Institute of Technology, Department of Computer and Information Science, Linköping University (2006)
6. Gamrad, D.: Modeling, simulation, and realization of cognitive technical systems. PhD thesis, Fakultät für Ingenieurwissenschaften, Abteilung Maschinenbau und Verfahrenstechnik der Universität Duisburg-Essen (2011)
7. Goetze, S., Moritz, N., Appell, J.E., Meis, M., Bartsch, C., Bitzer, J.: Acoustic user interfaces for ambient-assisted living technologies. Inform. Health Soc. Care **35**(3–4), 125–143 (2010)
8. Herz, J.: Kalorio (2015). http://www.kalorio.de. Accessed 27 Feb 2015
9. IBM, Institute of Culinary Education: Cognitive Cooking with Chef Watson: Recipes for Innovation from IBM & the Institute of Culinary Education. Sourcebooks (2015)
10. Larsson, S., Traum, D.: Information state and dialogue management in the TRINDI dialogue move engine toolkit. Nat. Lang. Eng. **6**, 323–340 (2000)
11. Larsson, S., Villing, J.: The DICO project: A multimodal menu-based in-vehicle dialogue system. In: Proceedings of the 7th International Workshop on Computational Semantics, IWCS-7, pp. 351–354, Tilburg, The Netherlands (2007)
12. Lison, P.: Structured probabilistic modelling for dialogue management. PhD thesis, Department of Informatics, University of Oslo (2014)
13. Morbini, F., Audhkhasi, K., Sagae, K., Artstein, R., Can, D., Georgiou, P., Narayanan, S., Leuski, A., Traum, D.: Which ASR should I choose for my dialogue system? In: Proceedings of the SIGDIAL 2013, Metz, France, 22–24 August 2013, pp. 394–403 (2013)
14. OpenDial: Opendial toolkit (2015). http://www.opendial-toolkit.net. Accessed 9 July 2015
15. Rennies, J., Goetze, S., Appell, J.E.: Human-centered design of e-health technologies: concepts, methods and applications, IGI Global, chap Personalized Acoustic Interfaces for Human-Computer Interaction, pp. 180–207 (2011)
16. Schäfer, U., Arnold, F., Ostermann, S., Reifers, S.: Ingredients and recipe for a robust mobile speech-enabled cooking assistant for german. In: KI 2013: Advances in Artificial Intelligence, no. 8077 in Lecture Notes in Computer Science (LNCS), pp. 212–223. Springer (2013)
17. SousChef: Cook Something Amazing Today (2015). http://www.acaciatreesoftware.com/. Accessed 9 July 2015
18. Williams, J.D.: The best of both worlds: unifying conventional dialog systems and POMDPs. In: Proceedings of the 9th Intern Speech Communication Association (Interspeech 2008), Brisbane, Australia, pp. 1173–1176 (2008)
19. Wolf, K.I., Goetze, S., Wallhoff, F.: Cooco, what can I cook today? In: Kendall-Morwick, J. (ed.) International Conference on Case-Based Reasoning ICCBR 2015, Workshop Proceedings, pp. 229–236(2015)
20. Wolf, K.I., Goetze, S., Wellmann, J., Winneke, A., Wallhoff, F.: Concept of a nutrition consultant application with context based speech recognition. In: Proceedings of the Workshop Kognitive Systeme 2015, Bielefeld (2015)
21. Zbeida, M., Goldsmitz, R., Shimony, T., Vardi, H., Naggan, L., Shahar, D.R.: Mediterranean diet and functional indicators among older adults in non-mediterranean and Mediterranean countries. J. Nutr. Health Aging **18**(4), 411–418 (2014). Springer

Learning Behavioural Routines for Early Detection of Health Changes

Raoul Hoffmann, Axel Steinhage and Christl Lauterbach

Abstract A persons daily routine is a valuable early indicator of a changing health status. Ageing diseases in an early stage have an impact on measurable behaviours like sleeping times and quality, consistency of activity sequences and gait characteristics. Unfortunately, these interesting parameters of daily routine are hard to assess. In this work a technical system is proposed that is capable of detecting those changes in indoor daily routine automatically, unobtrusively and over long periods of time. The system relies on a combination of various sensors, including a novel capacitance sensing system covering large areas in the home environment, that is capable of detecting the locations of persons. A processing scheme is proposed to extract behavioural information from the sensor data, which is fed into a learning algorithm that internally represents typical patterns, and outputs a measure for the divergence of current behaviour from typical behaviour. As a noticeable feature, no interaction with the patient is required.

1 Motivation

As people get older, their need of medical care often increases. In the meanwhile, the ratio of carers to patients decreases due to the well-known demographic changes. Both effects combined result in the necessity to keep people healthy and in an autonomous lifestyle as long as possible, and deliver the right amount of care considering a person's health status. Besides economic advantages, personal benefits result from a wisely chosen amount of care. A too low level of medical treatment can result

R. Hoffmann (✉) · A. Steinhage · C. Lauterbach
Future-Shape GmbH, Altlaufstr. 34, 85635 Höhenkirchen-Siegertsbrunn, Germany
e-mail: raoul.hoffmann@future-shape.com

A. Steinhage
e-mail: axel.steinhage@future-shape.com

C. Lauterbach
e-mail: christl.lauterbach@future-shape.com

R. Wichert and B. Mand (eds.), *Ambient Assisted Living*,
Advanced Technologies and Societal Change,
DOI 10.1007/978-3-319-52322-4_12

in faster physical deterioration, while a too high level of medical attention by carers and doctors can be considered a nuisance.

Currently, someone's health status is mainly diagnosed during doctoral visits. The system proposed in this work aims at indicating a change in behaviour as a possible result of a newly developed disease. Autonomous sensors and systems that are embedded ambiently in the common environment can extend appointment-wise medical measurements with this new kind of information source. Typical tools of behavioural diagnosis like questionnaires are possibly imprecise and can inevitably only be carried out with low frequency, furthermore they pose an inconvenience to the patients if they need to be filled out regularly. Compared to this, technical systems can run continuously and deliver measurements of finer granularity without disturbing anyone.

2 Sensor Floor for Behaviour Monitoring

SensFloor is a sensor system that is installed concealed under the floor in large areas in indoor environments [1, 2]. The sensor is capable of tracking the locations of persons and pets that move on it, as well as detecting other events like a person that has fallen and now lies on the floor.

SensFloor contains independently operating electronic modules, that continuously perform measurements of the electric capacitance in eight discrete triangle-shaped areas. The electronic modules of the sensor are flexible circuit boards with a MCU and on-board data radio which are embedded in a textile underlay. The underlay is a polyester fleece base with a further textile layer on top that is metallically coated and hereby made electrically conductive. As seen in Fig. 1, patterns are cut

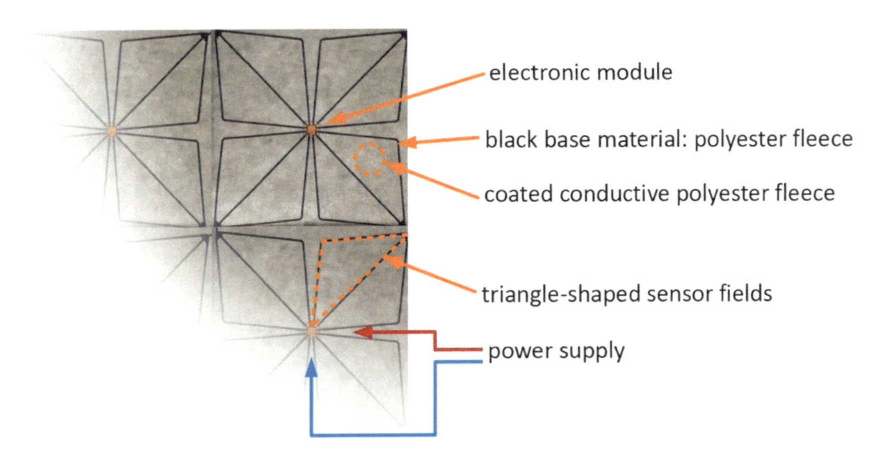

Fig. 1 Floors of large size can be covered with these sensor modules

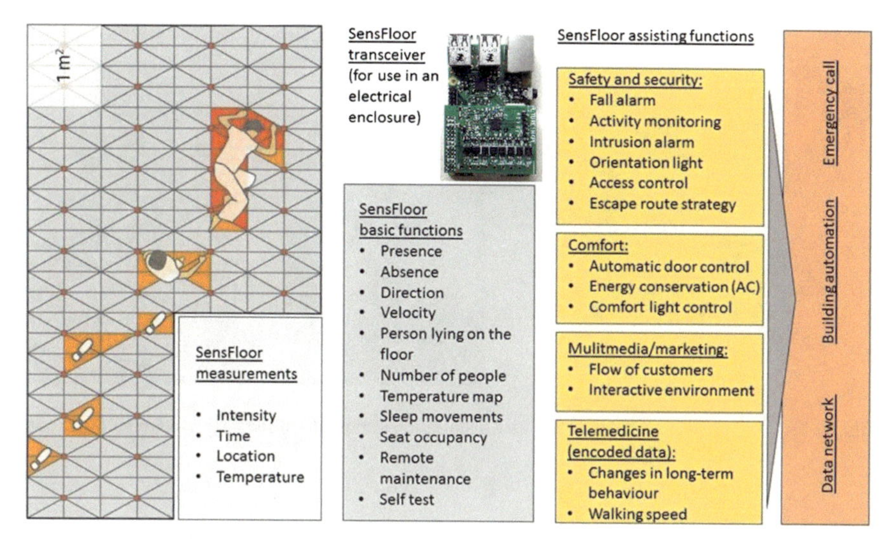

Fig. 2 Overview of SensFloor applications

into the top layer. This is used to form the triangular shaped individual measurement areas, as well as lines used to supply the modules with power.

The modules send out the results of the capacitance measurements wirelessly to a central transceiver. Large areas can be covered with the sensor by laying out and connecting sensor reels next to each other, and applying power at the edge of the area. SensFloor can be used with a wide variety of typical floorings, like parquet, carpet, tiles or PVC. When covered with flooring, the sensor is invisible to the inhabitants and therefore unobtrusive.

An application of SensFloor in care is the prevention of falls by switching on lights at night when a person leaves the bed, helping elderly people to orientate themselves [3]. In case a fall happens nevertheless, it can be detected, and help can be called automatically. Other applications are found in the area of Smart Homes, whenever locations of persons need to be determined. SensFloor functions and applications are shown in Fig. 2.

3 Experimental Setup

Within the BMBF-funded CogAge project (FKZ16SV7311), the proposed system is installed in an apartment in the city of Siegen in Germany. Figure 3 shows the floor plan and a photo of the apartment. After installation, the apartment is inhabited, so that live data from a person's daily routine can be recorded over a long period of time. The inhabitants agree to have behavioural data collected and evaluated for the purpose of validating learning algorithms that perform daily routine prediction

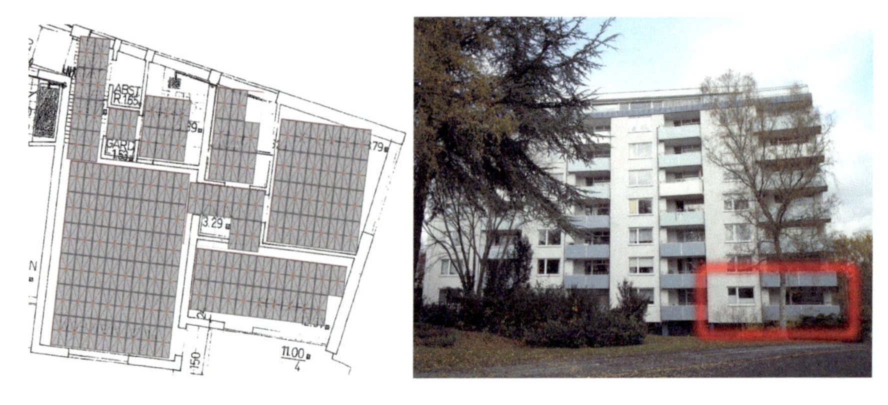

Fig. 3 Floor plan with projected SensFloor installation and photo of the CogAge Demo apartment

and detection of variations. The apartment is equipped with SensFloor in all areas and further sensors. Data from all sensors is gathered locally in the flat, and can be accessed over the internet using a secure Virtual Private Network (VPN) tunnel. The collected data is then used to develop and test algorithms that learn about the life of the person living there.

An important and beneficial property of the sensor floor is, that the data does not reveal the identity of a person. In contrast to camera sensors, SensFloor can be used in areas where preserving intimacy is highly important, for example in bathrooms. Generally, ambient sensors that do not collect identity information experience a higher acceptance rate by people, compared to typical surveillance sensors like cameras.

4 Learning Behavioural Routines

Raw data from the sensors in the demo apartment is interpreted and described as states of semantic meaning, that serve as input to the learning system. These states are assumptions about a person's activity based on their presence in areas in their home that serve a specific purpose. Examples of such activities of daily routine are the following:

- Entering and leaving certain rooms in the apartment
- Entering and leaving the apartment itself
- Getting up and going to sleep (estimation of sleep quality and behaviour)
- Presence in the kitchen and dining area (eating/drinking regularly and enough amounts)
- Presence in the bathroom (personal hygiene)
- Presence in the living area (entertainment and intellectual activities)

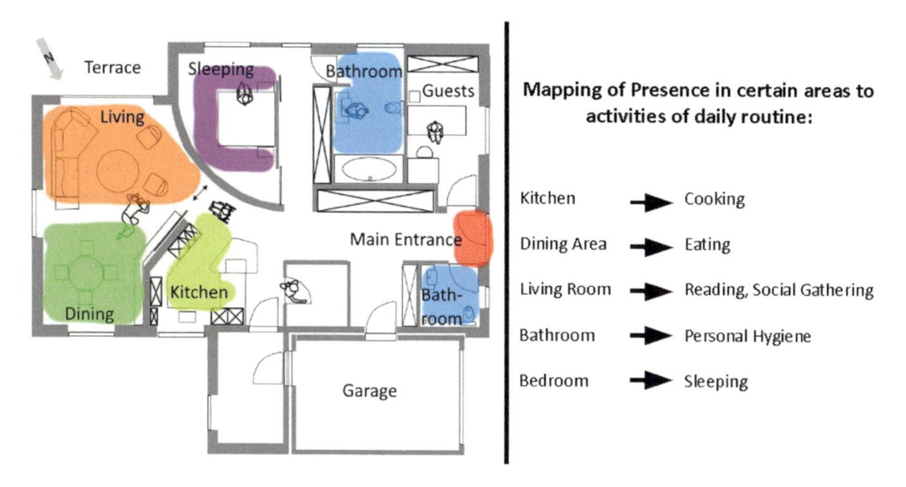

Fig. 4 Example of a floorplan with areas designated for typical daily activities

The task of the algorithms developed in this project is to take these semantic signals as input and learn certain characteristics about them. One important characteristic is the consistency of the sequence of events, meaning: Some activities are in normal cases followed by other activities [4]. As an example, getting up in the morning is typically followed by a sequence of activities that depends on how the individual person is used to do things. A common case of a sequence is to get up, walk into the bathroom, stay there for some time, go into the kitchen to prepare coffee and get things to eat, and then go to the dining table to have breakfast.

Phases of irregularity in such a typical routine could indicate a beginning mental problem, for example an elderly person forgetting to have breakfast in the morning. In such a case, some increased attention by caring personnel could prove valuable. Another interesting characteristic is the frequency and time dependency of activities. An example for this would be the times of getting to bed and up again. An algorithm that learned that usually no such events happen in the middle of the night, should trigger a health state change indication when a person suddenly starts getting up several times in the middle of the night, suggesting some sort of sleep disorder needing treatment (Fig. 4).

Besides information that can be linked to actual activities of daily life, the SensFloor data can possibly be filtered to get further information about the physical activity persons perform in their home. Examples of such measurements are the walking distances per day, the walking speed and step count. Furthermore, it is planned to extract parameters describing the gait from the SensFloor data. Some of these features have the potential to be medically relevant indicators, like stride length variability and frequency, tendencies to tumble, drag a feet or walk with a limp [5–7]. An algorithm that is capable of performing this information processing task needs to be an adaptive dynamical system that can learn the regular behaviour of an individual person. Behaviour that can be considered normal varies vastly between individuals. The computing system will have to learn the individual behaviour over some time, and then report any changes in the behaviour patterns that occur later on. A class of algorithms that have shown to be practical in such a use case are Artificial Neural Networks (ANNs). ANNs are inspired by research on natural computing systems in biology, as for example the structure and computational principles of nervous systems in the human brain. Research on ANNs resulted in many different kinds of structures for various tasks, from classification and regression problems to time series predictions. A common property of ANNs is that they are capable of learning from input signals, without the need to specify and provide a model of the observed system for which the dynamics should be learned.

A species of ANN that was used successfully in predicting time series is the "Liquid State Machine" (LSM), a kind of neural network consisting of a randomly connected reservoir of artificial neuronal computing units that are connected to a structured readout layer performing conventional supervised learning strategies [8, 9]. A similar approach is possibly applicable in the setup described here. The resulting algorithm takes the input signals extracted from the sensor floor as input, and predict the future signal course based on the learned representation of patterns. Changes in the patterns can be detected by comparing the predicted signals to the actually occurring signals.

5 Conclusion

We proposed a system that is a combination of a sensor floor installed in the home of a person in need of medical care, and a fitting analysing algorithm. Signals from the sensor are evaluated to extract information about activities of daily living and physical parameters describing gait characteristics, and fed as inputs into the learning system. The algorithm needs to build up an internal representation of the observed persons daily routine, to perform time series prediction of activity patterns, and further report any changes in these patterns. The system is used to indicate the current level of divergence from routine, to notify carers of the possibility of a changed health status.

Acknowledgements This work is supported by the German Federal Ministry of Education and Research (Bundesministerium für Bildung und Forschung, BMBF) within the project "Cognitive VillAge: Adaptiv-lernende, technische Alltagsbegleiter im Alter" (CogAge, FKZ 16SV7311), in which research of adaptively learning technology supporting elderly persons in their everyday life is pursued.

References

1. Steinhage, A., Lauterbach, C.: SensFloor and NaviFloor: large-area sensor systems beneath your feet. In: Mastrogiovanni, F., Chong, N.-Y. (eds.) Handbook of Research on Ambient Intelligence and Smart Environments, Trends and Perspectives. IGI Global Hershey, PA (2011)
2. Steinhage, A., Hoffmann, R., Lauterbach, C.: Automatische Unterscheidung von Personen und Haustieren auf dem Assistenzsystem SensFloor. In: AAL-Kongress 2015. VDE VERLAG GmbH (2015)
3. Lauterbach, C., Steinhage, A., Techmer, A.: A large-area sensor system underneath the floor for ambient assisted living applications. In: Pervasive and Mobile Sensing and Computing for Healthcare, pp. 69–87. Springer (2013)
4. Fleury, A., Vacher, M., Noury, N.: SVM-based multimodal classification of activities of daily living in health smart homes: sensors, algorithms, and first experimental results. IEEE Trans. Inf. Technol. Biomed. **14**(2), 274–283 (2010)
5. Verghese, J., Lipton, R.B., Hall, C.B., Kuslansky, G., Katz, M.J., Buschke, H.: Abnormality of gait as a predictor of non-Alzheimers dementia. New Engl. J. Med. **347**(22), 17611768 (2002)
6. Beauchet, O., Allali, G., Annweiler, C., Bridenbaugh, S., Assal, F., Kressig, R.W., Herrmann, F.R.: Gait variability among healthy adults: low and high stride-to-stride variability are both a reflection of gait stability. Gerontology **55**(6), 702–706 (2009)
7. Wittwer, J.E., Webster, K.E., Andrews, P.T., Menz, H.B.: Test retest reliability of spatial and temporal gait parameters of people with Alzheimers disease. Gait Posture **28**(3), 392396 (2008)
8. Maass, W., Natschläger, T., Markram, H.: Real-time computing without stable states: a new framework for neural computation based on perturbations. Neural Comput. **14**(11), 2531–2560 (2002)
9. Burgsteiner, H., Kröll, M., Leopold, A., Steinbauer, G.: Movement prediction from real-world images using a liquid state machine. Appl. Intell. **26**(2), 99–109 (2007)

Enabling an Internet of Things Framework for Ambient Assisted Living

Helmi Ben Hmida and Andreas Braun

Abstract Ambient Assisted Living (AAL) technologies hold great potential to meet the challenges of health, support, comfort and social services in European countries. After years of research, innovation and development in the field of health care and life support, there is still a lack of good practices on how to improve the market uptake of AAL solutions, how to commercialize laboratory results and prototypes and achieve widely accepted mature solutions with a significant footprint in the European market. The Internet of Things (IoT) consists of Internet connected objects such as sensors and actuators, as well as Smart appliances. Due to its characteristics, requirement and impact on real life system, the IoT has gained significant attention over the last few years. The major goal of this paper is to strategically specify and demonstrate the impact of the usage of IoT technology and the respect of IoT specification on the quality and future collaborative usage and extendability of deployed AAL solutions in real life.

Keywords Internet of Things · Enabling ambient assisted living · AAL for real life · New generation for AAL solutions

1 Introduction

Ambient Assisted Living (AAL) technologies hold great potential to meet the challenges of health, support, luxury and social services in European countries. In fact, supporting people in independent living using Information Communication Technology (ICT) is a great opportunity to solve the upcoming challenges. Different ICT as well as technical components, such as sensors or actuators are integrated in one

H. Ben Hmida (✉) · A. Braun (✉)
Fraunhofer Institute for Computer Graphics Research IGD, Fraunhoferstr. 5,
64283 Darmstadt, Germany
e-mail: helmi.ben.hmida@igd.fraunhofer.de

A. Braun
e-mail: andreas.braun@igd.fraunhofer.de

R. Wichert and B. Mand (eds.), *Ambient Assisted Living*,
Advanced Technologies and Societal Change,
DOI 10.1007/978-3-319-52322-4_13

system in order to make this possible used autonomously. The users can use it to participate in social and professional life more easily, secure, and independently. Even after years of research, innovation, and development in the field of health care and life support, there is still a lack of good practices on how to improve the market uptake of AAL solutions, how to commercialize laboratory results and prototypes and achieve widely accepted, mature solutions with a significant footprint in the European market. There is a chance to make Europe the largest region to adopt and align AAL and innovations from related fields.

From another side, the Internet of Things (IoT) is covering various aspects of the extension of the Internet and the web in the physical world, through widely spatially distributed devices with embedded identification, sensing capabilities and intelligent acting. In this perspective, the IoT will be more things-oriented, thus managing, deploying and coordinating in an intelligent way with sets of smart objects. This innovation will be enabled through the inclusion of electronics in everyday physical objects, making them smart and allowing them to seamlessly integrate into the global infrastructure of the physical Internet. This will lead to new opportunities for the ICT sector, paving the way for new services and applications are able to take advantage of the link between the physical and virtual worlds [1].

The major goal of this paper is to strategically specify and demonstrate the impact of the usage of IoT technology, the respect of IoT philosophy and system flexibility on the quality and future collaborative usage and extendability of deployed AAL system in real life. With respect to the previous vision, we must attire the reader attention that this paper will not address the impact of IoT on AAL from a theoretical laboratory perspective, but from real deployed AAL systems in real life one and therefore fitting the real life requirement, especially related to system extension, interoperability, personalisation and adaptation. Although the diversity and the high quality of the available AAL solutions, they have a main common factor which is their closed aspect [2]. Closed platforms are a software system, where only the service developers and/or provider have access and control over the applications. Closed systems are only accessible for specific, homogeneous and compatible applications or devices. Actually, most AAL products and technologies are coming as closed black box, where the end user can simply profit from the presented service as its original form [3]. Similarly, only the original developer is able to update and change the source code. From the point of view of connected devices, only pre-installed sensors and actuators can be taken into consideration. No new functionalities, sensors, or actuators can be straightforwardly added to the original setup [4].

The following paper is structured into Sect. 2 which gives an overview about the IoT domain, architecture, advantages and related requirement. Section 3 discuss and argue the feasibility and foresee the added value behind the usage of IoT architecture for AAL solutions. Section 4 present the created AAL solution based on IoT system requirement. Section 5 perceives the impact from a real life perspective of the created solution. Finally, Sect. 6 concludes the paper with the learned lessons and the future directions.

2 Advantages of an IoT Architecture and Derived Requirements

Over the past decade, the field of IoT has gained much attention especially in the industry area due the technical operational capacity such a field might offer [5]. It aims to create a world where everything as intelligent objects are linked to the Internet and communicate with each other with a minimum of human intervention. The ultimate goal is to create "a better world for humans" where things around us know what we like, what we want, what we need and act accordingly without explicit instructions, which is well aligned with the ubiquitous computing philosophy.

The term IoT is used to cover the various aspects related to the extension of the network and the Internet into the physical world, through the deployment of a wide range of devices that are distributed spatially and typically integrate sensing, processing, and acting. As IoT is still in early development phase [6], the research there is still in its infancy [7]. Therefore, there are no standard definitions for technology operations. The following definitions are provided by different researchers.

- Definition by Tan et al. [8]: Things have identities and virtual personalities operating in smart spaces using intelligent interfaces to connect and communicate within social, environment, and user contexts.
- Definition by the European Commission [9]: The semantic origin of the expression is composed by two words and concepts: Internet and Thing, where Internet can be defined as the world-wide network of computer networks, based on a standard communication protocol, the Internet suite (TCP/IP), while Thing is an object not precisely identifiable Therefore, semantically, Internet of Things means a world-wide network of interconnected objects uniquely addressable, based on standard communication protocols.
- Definition by Saint-Exupery [10]: The Internet of Things allows people and things to be connected Anytime, Anyplace, with Anything and Anyone, ideally using Any path/network and Any service.

From an AAL perspective, the first definition seems to be the most suitable to enable the IoT for AAL. This definition encapsulates the AAL vision, especially related to devices, environment, context, and Ubiquitous Computing [11].

2.1 *The Essential Characteristics of an IoT System*

In this section, we will briefly highlight the characteristics of an IoT system. This is necessary, in order to specify the main criterion that an AAL system must fulfill in real-life to be an IoT compatible system.

- Intelligence: Where a context awareness framework must be part of the installed and deployed IOT-AAL system. It aims to Collect and transform the Row data to knowledge and reason about it in an intelligent way.

- Architecture: Internet of things architecture, with reference to [12, 13], should make uses of a mixed architecture. Primarily there would be two architectures: event driven and time driven. Some sensors produce data when an event occurs (e.g. door sensor, motion sensors); the rest produce data continuously, based on specified time frames (e.g. environment sensor). Mostly, IoT are event driven where common rules are used in such systems.
- Complex system: The IoT architecture is composed of a big number of objects (sensors and actuators) that must autonomously interact with the corresponding environment and the related users. Newly added devices must be automatically recognized, added, and start providing their functionalities. The processing and interaction with the different things will depends on the added value provided by the different devices. In some cases, passive things will simply provide measurements. In other cases, objects may have larger memory, better processing, and advanced reasoning capabilities, which make them more intelligent.
- Size considerations: It is predicted that there will be 50–100 billion devices connected to the Internet by 2020 [14]. The IoT framework and the related system architecture need to take such a complexity in consideration and especially to ensure the interaction among these objects.
- Time considerations: The real-time processing of the device data, the real-time recognition of several contextual situations, and the related decision must be ensured by IoT systems.
- Service-oriented architecture: Due to the requirement of an IoT system to be a distributed, flexible system, consuming resources as a service has become the most common approach [15]. Everything-as-a-service, especially based on platform is highly efficient, scalable, and easy to use [16].

2.2 Essential Requirements of an IoT System

From a system level point of view, it can be seen that the Internet of things as a distributed system of things consisting of a very large number of smart appliances as producers and consumers of information. The ability to interface with the physical word is achieved through the presence of devices able to sense physical phenomena and translate them into a stream of information, or through other devices able to receive conceptual information and translate it onto actions.

From a service level perspective, creating, composing and integrating services or resources provided by smart objects, are the main challenges that must be addressed [17]. This requires the definition of an architectures and methods for virtualizing objects. This will requires the usage of a standardized representation of smart objects, thus hindering the heterogeneity of devices/resources. Similarly, it will require making use of methods for flawlessly integrating and composing the resources and services of smart objects into value-added services for users.

Finally, from a conceptual level, the IoT architecture and especially applications builds on three main pillars, related to the ability of smart objects to be identifiable, interactive and communicative.

2.3 The Paradigm of IoT and System Architecture

The Internet of Things was originally thought as extending the principles of the Internet as a network organisation concept to physical things. That is, things would get a unique ID, which is machine readable and an associated digital representation on the web [18]. Whereas ubiquitous computing was designed to make objects intelligent and create richer interaction, the Internet of Things was much more focused on virtual representations of automatically identifiable objects. Obviously, both concepts are important pieces of a future Internet and the IoT scope. In this regard, the IoT applications will requires interaction with heterogeneous sensors, aggregators, actuators, and a diverse domain of context-aware applications, while preserving the security and privacy. In order to meet the above requirements, the IoT will require the usage of a software platform, in this context called middleware, ensuring the abstraction to applications from the things, and offering multiple services. There have been a lot of researches towards building up this middleware addressing interoperability across heterogeneous devices serving diverse domains of applications, adaptation, context awareness, device discovery and management, scalability, managing a large data volumes and, privacy, security aspects of the IoT environment [19]. In fact, the middleware for IoT is required as it is very difficult to define and enforce a common standard among the diverse devices belonging to the different domains in IoT. Therefore, a middleware acts as a bond, joining the heterogeneous components together, providing application programming interfacing (API) for the physical layers and required services to the applications, thus hiding all the details of diversity. A middleware will also ensure the unification of the sensing framework, a formal modeling and representation of the real world, pluggable reasoning engines for high-level contexts recognition and delivery of runtime service composition mechanisms thus simplifying and empowering the development of context aware applications.

3 IoT and AAL

We must recall that the paper scope is not a specification or a contribution to the IoT domain, but mainly the extraction and synthesise of the major characteristics, requirements and best practice that IoT systems are using, thus enabling it for a successful deployment of AAL systems that link to the IoT ones.

AAL builds upon Ambient Intelligent, whereas the latter one aims at building digital environments that are aware of the humans presence, their behaviors and needs through a very rich in sensing, computing and actuation capabilities that are designed to respond in an intelligent way to changes in the environment and the user needs, in order to support them carrying out specific tasks. It builds upon the Ubiquitous Computing concept [20]. Aarts et al. [21] summarise the five key features that characterise an AmI system:

- Embedded: Networked devices are integrated into the environment.
- Context aware: System recognises the environment, related embedded devises, people and their situational context.
- Personalized: System can tailor itself to meet peoples needs and profile.
- Adaptive: System can change in response to people and environment changes.
- Anticipatory: System anticipates peoples desires without conscious mediation.

Compared with the highlighted requirement of an IoT system in Sect. 2.1, we can conclude that the Context aware, the Personalized, the Adaptative and the Anticipatory feature meets the intelligence requirement of IoT while the Embedded one corresponds the IoT complex system architecture.

Aligned with the previous analysis, Wichert et al. [22] stated that AAL applications area shares several factor with the IoT ones. Most AmI systems are based on the inclusion of sensing and computing capabilities embedded in the environment, where context-awareness is the key element for reasoning, both in AmI and IoT. Accordingly, one of the main focuses of research in AmI has been the development of reasoning techniques for inferring activities of users and devising appropriate response strategies from the embedded devices [23]. Due to the highlighted intersections, the IoT system architecture and specification with its awareness about real life system constraint and major facilities will enable the following concepts for AAL:

- Concept of integration of open scenarios, whereby new functions, capabilities and services need to be accommodated at run-time, without them having been necessarily considered at design time.
- Concept of self-configuration and self-organization, possibly cognitive, capabilities needed to provide this additional degree of flexibility.
- Concept of loosening coupling: a loosing link between devices, system, reasoners and interfaces will enable Plug-and-Play of devices, system flexibility and interoperability between the different component.
- Concept of system distribution: The assisted person environment is not restricted to specific location (e.g. home), but might be extended to his car, hospital, green places… where linking the different distributed appliances will be with a major added value for a full awareness of the end user context and situation. Same requirement will fit the relation between different systems like the a local (e.g. apartment) and a more global one (e.g. building).
- Concept of adaptability to diverse contexts, with different resources available and possibly deployment environments changing over time.
- Concept of interoperability between the different system components.

With reference to the specified concepts and driven by the main advantages the IoT might bring real life AAL system in term of intelligence, flexibility, adaptability and distributions… the next section will highlight our contribution in terms of an IoT-enabled AAL system that has already been deployed in real life, showcasing the advantages and providing lessons learned.

4 uSmAAL: A Real-Life Use Case for an IoT-Enabled AAL System

Supporting people in independent living using ICT is a great opportunity to solve upcoming societal challenges. Primarily these are related to the ageing of the European population where AAL solutions aims to increase the quality-of-life that causes increased demands on comfort and luxury [24–26]. AAL technologies, Smart Homes, Smart Buildings and eHealth applications that are based on advanced IoT technologies may help in increasing peoples security, support, comfort and therefore reduce the consumption of resources associated to buildings (electricity, water). It can also improve the satisfaction level of inhabitants. Within the created system, a key role is played by sensors and actuators which are used to both monitor tenant security, increase comfort, optimize resource consumption as well as to proactively detecting the current needs of the users. Such a scenario integrates a number of different subsystems and hence requires a high level interoperability add to the ability to reason in a distributed and cooperative way to control the environment in order to ensure that decisions taken on the resources under control (e.g. switch on/off lighting, heating, cooling, etc.) are in line with the users needs and expectations. They are strictly intertwined to current or future activities of the tenants. Based on the prerequisites and collected requirements from one side, and the main gathered advantages of the IoT field in real life, the universAAL Smart AAL System (uSmAAL) has been created. It forms an open, flexible, reusable and easily expandable system for providing smart AAL services based on the open source platform universAAL [27]. The created AAL system follows the IoT architecture and incorporates the desired robustness, flexibility and extendability into its system architecture. The created system is running on top of the semantic open platform universAAL and is mainly composed of a middleware, a context awareness module, layers for flexible access, abstraction, and integration, and a set of sensors and actuators. The architecture is shown in Fig. 1.

4.1 Sensing/Acting Layer

From the IoT devices perspectives, all devises are characterized by a unique identifier and unique address mainly composed by the IP address of the related Gateway/controller add to the local device address. The different devices have different locations, characterized by different addresses and connection protocols (light, heater, presence sensors, phones, blood pressure…). Currently, the implemented

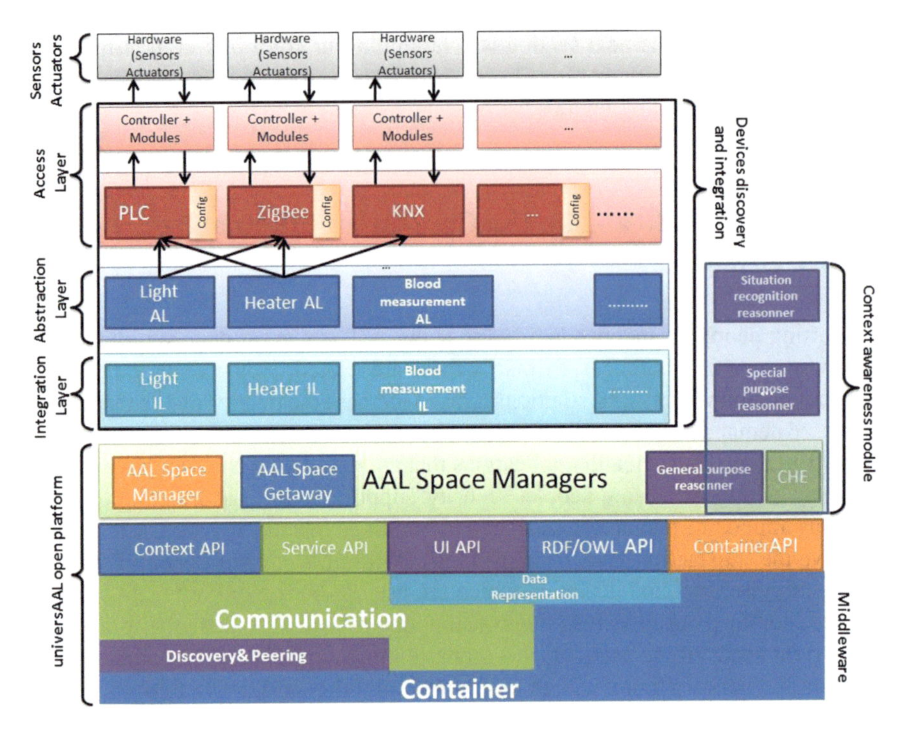

Fig. 1 uSmAAL system architecture

system cover more than 1000 devices distributed in the different locations the system is covering. Its mainly composed by two components: actuators and sensors. The actuators represent hardware devices which can be controlled, e.g. lights, heaters, outlets, blinds, etc. This means it is possible to change the status of the device, by performing some interaction. This interaction can be performed by direct change of the actuator values, or automatically by the system itself (e.g. reasoning). In both cases the controller will perform an action which will result in the desired outcome. It may also be possible that an actuator is controlled manually like switching a lamp by pressing a button. If so, the controller will be automatically notified about such an action. The second component is represented through the sensors. A sensor measures values and registers modifications in the environment and notifies the controller about it. A sensor can before example a motion sensor, a light sensor, a temperature sensor or a CO_2 sensor… Since sensors are measuring specific variables of the environment (temperature, CO_2 value, etc.), they can keep track of the actual status of it. Thus they build the basis for specified rules and reasoning. For example reasoning like turning on the light in a room if there is movement at night, is based on the sensors for movement and daylight.

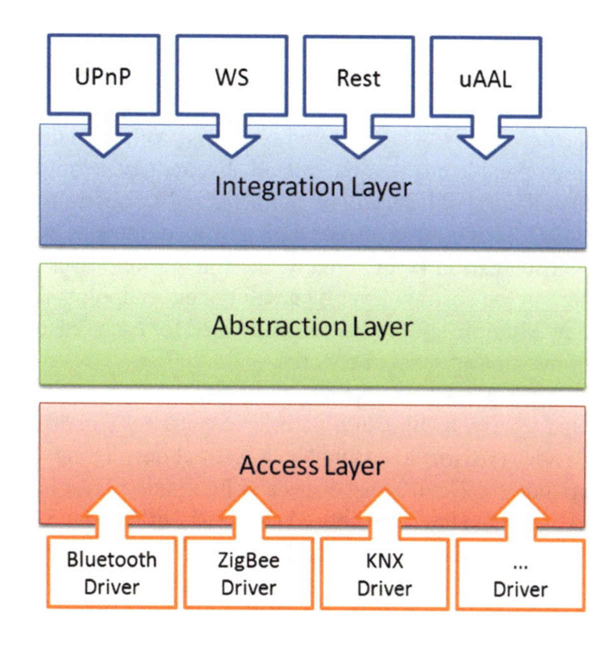

Fig. 2 UniversAAL device discovery and integration structure

4.2 *Device Discovery and Integration Layer*

The universAAL Device Discovery and Integration is an abstraction layer able to represent and to facilitate the integration of sensors and actuators in the platform. From the analysis of the solutions adopted for the different network technologies, a common pattern is emerged as far as the architecture layering is concerned for the sensor and actuator networks integration. The created architecture is typically organized upon three sub-layers namely the Access, Abstraction and Integration, Fig. 2.

Access Layer: The Access Layer is usually composed of a number of technology drivers, sometime already integrated in the operating system with well-defined API. There are two approaches generally followed to use the drivers API. In case of vertical solutions, a controller interacts directly with the specific remote devices through the drivers and sometime provides a user interface to configure the device. The creation of proxies can be automatized whether the technology supports a discovery mechanism, and notification of new devices joining/leaving the sub-network is provided. A controller or a gateway specifies a standard for the communication where different hardware devices, mainly in a smart environment, are connected to. Within the created AAL system, the access layer aims to establish the communication between the created IoT AAL Device abstraction layer and the concrete IoT appliances. Examples of implemented protocols are PLCs (Programmable Logic Controller) [28] KNX controller [29] or ZigBee controller. [30].

Abstraction Layer: The Abstraction layer aims to transforms the proxies, which are a flat and raw representation of the physical networked nodes installed in the Internet to more significant device objects, with an easier and intuitive interface for the developers. As the created system is profit from the semantic capability as a main interoperable enabler, the related abstract device ontologies are also defined, where the different objects will be characterized by different Uniform Resources Identifiers (URI). This layer is also occupied by a component integrator and discovery.

Integration layer: Finally, the Integration Layer publishes the proxies instantiated in the abstract layer by creating new endpoints according to a specific technology. Multiple end-points can be created for each sensor, thus providing multiple way of integrating sensor networks.

The three layers whether implemented on the same host machine or in a distributed system architecture describe a gateway solution adopted to map an IP-based devices (sensors and actuator) to virtual one. Through the installed uSmAAL AAL system, the Device Discovery and Integration layer has been implemented and aims to builds the interface between the real world and the Abstraction level of uSmAAL. This layer receives commands from the created AAL services and applications and translates them into low level ones depending on the addressed devices. This means if a specific device is turned on by some an AAL service, the layer will receives this command and redirect it to the correspondent device which, based on the hardware protocol, transfer the command to the addressed device. The other way around, if a device is manually turned on, the AAL service will be automatically notified, same in case of new devices. As this layer builds the bridge between the virtual word and the real one, it represents a core component of the system since it allow a complete separation between the Real word and the virtual one.

4.3 Context Awareness Module

The uSmAAL system is classified as a modular architecture, where its composed by several modules dynamically communicating to each others. The modular approach, as a key for future programming technology, aims to divide the set of complete application to basic modules. It intends to facilitate the re-combination of modules differently, thus offering several new sets of services. The recombination process is reinforced through several components developed within the universAAL system, mainly the context awareness framework. As previously discussed, context awareness plays a crucial role in any ubiquitous computing, AmI and AAL systems. The main goal of the context awareness is to improve the usability of applications through adapting their comportment dependent on the context. Ambient assisted living aims to anticipate user needs, according to the situation they are in, by means of semantically understanding the environment status and on-the-fly composition of different service components. This requires applications to be able to understand the context and situation the user is in. Such a theme has been addressed within the ambient intelligence, ambient assisted living and pervasive computing fields, leading to a

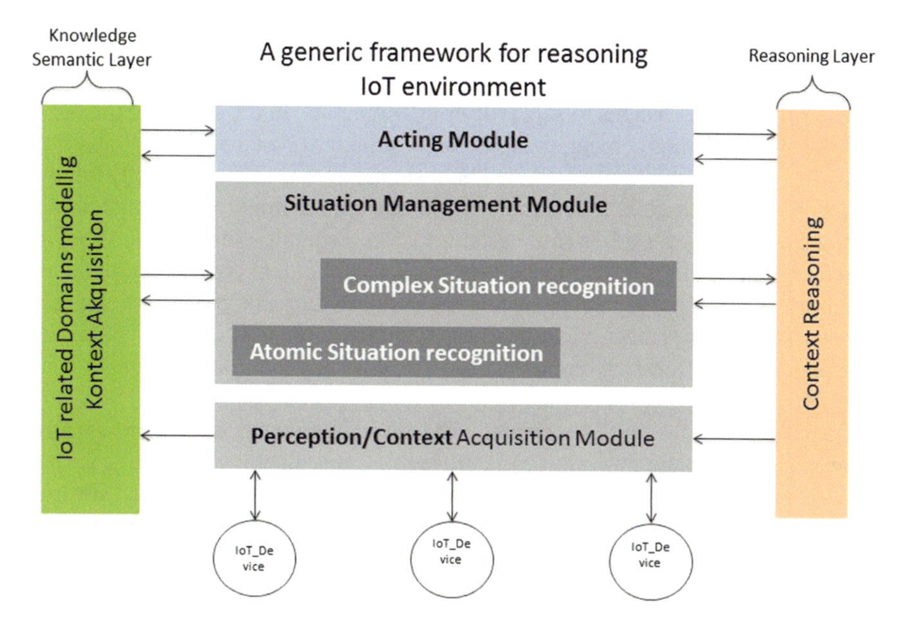

Fig. 3 A generic framework for reasoning IoT environment

number of solutions able to leverage contextual information coming from a number of sources. As observed till now, The created system is also enriched by a context awareness module able to semantically perceive the virtual device status.

The created module aims implement and extend the universAAL context awareness framework [31] to create new intelligent rules recognizing high level semantic context (situation) or a combination of AAL services to be executed under certain recognized circumstances. Figure 3 present the main component collaborating togethers thus ensuring a very flexible and complete context aware module for IoT.

IoT Related Domains Modelling: The created module aims to provides the capabilities for powerful high level reasoning based on semantic web technology. To do, a main first step consist on the semantic modelling of contextual data related to the addressed environment hardware. Also, as the IoT related application require a kind of virtualization of the different connected hardware to maximize its usage in real life situation, the created system has made an attempt to abstract and semantically model the whole environment with all its details, especially through the implementation of the following ontologies: Physical World, Profile, Furniture, Device, Service, Context and situation ontology…

Context Perception and Acquisition: The context perception and acquisition sub-module aims to instantiate the different created ontologies though the Devise discovery and integration layer, more specifically, the integration layer. This will held to the creation of a Knowledge base composed by the ontologies schema and the related instances.

Situation Management: Situation awareness can be seen as the capability of the different created entities to be aware of the situation changes and automatically adapt themselves to such changes to satisfy user requirements. In this concern, the Situation Management aims at recognizing different situations based on the available context. The different situation are divided onto simple ones if they are directly inferred based on specific context, for example an alarm situation might be recognized once a smoke context is created. In more advanced cases, complex situation like cooking activity might be inferred next to the combination of different context and/or simple situations. The continuing evolutions of the system in term of changes of the environment status makes the Situation Management one of the most desired IoT features to support the real life new circumstances thus ensuring the system adaptation and flexibility.

Semantic Reasoning: The created AAL system aims to reason on available knowledge, thus coming with more mature information about the air quality, the persons activities, the environment status, the energy consumption the medical situation of the users... The new inferred information will be therefore used for further decision (e.g. services execution, service combinations...). For example, one key technology installed behind the uSmAAL is the localization of residents in their homes and neighbourhoods. Once installed, the main recognized context will be related to the person position and posture. based on the basic context, more complex situation related the user activities will be inferred. At a third level, many services can be placed together such as modular fall detection, intrusion detection, alerts when leaving the house or if the window is still open; Warning if young children or people with dementia leave the bed, turning on the light when standing up from the bed, energy saving by turning of devices if the room is left and much more. Modular technologies will be the main key toward offering better services with minimum cost.

Acting Module The uSmAAL created system is full based on semantic modelling of the different entities. Among them, the different available services are also abstracted and semantically modelled. After succeeding modelling the different devices and situations, the provisioning of contextual information as a key functionality for the IoT system will enables through the acting module a context/situation based actions as the output of the Reasoner. The acting module will grant that the right service will be executed.

4.4 *UniversAAL Open Platform*

The above highlighted uSmAAL AAL function and services are running on the top of the semantic open platform universAAL. The latter one has been built in a community effort since 2010 and finally established in January 2014 [31]. It provides the result as an open source software platform for AAL domain. The universAAL platform is an open-source software platform for the development, operation and marketing of AAL applications. It is offered with the Apache Software-License 2.0 and is especially designed for the development of open and distributed AAL systems. As a semantic open platform, and thanks to its unique features [27], universAAL will allow the creation of a sustainable ecosystem of AAL applications and services providers. From developer point of view, open platform-universAAL offer an assisted development environment to create AAL application based on the universAAL open platform, a distribution of AAL services over different processing components in a network and sharing information and composing services between all developed application and services. It allows also the cross-exploitation of the offered functionalities, facilitating the extension of an application in terms of functionalities and devices and simplifying the adaptation of a certain application to new circumstances

5 Perceived Impact of IoT on the Deployed AAL System Architecture

The IoT, via its characteristics and requirement, has positively affected the created AAL system through changing the system architecture and increasing its flexibility and accessibility. More concretely, the system intelligence has been empowered through the inclusion of a context awareness module. In fact, transforming the Row data to knowledge and reason about it has increased the system flexibility, where the direct link between sensors and actuators has been avoided and alternated via an intermediate situations layer. Respecting the IoT characteristics has also empowered the system adaptability as all kinds of reasoning and acting are taking the end user and its current situation and profile in consideration and freely inferring the new actions or knowledge without any direct dependency with the hardware level.

As the IoT architecture support complex systems, the addressed Device Discovery and Integration layer has completely disconnected the high level AAL system, where all virtual devices and their related characteristics are created, from the low level one where the devices, their ip addresses and access protocol are defined. In case of AAL system for real life usage, the provision of the complex architecture has supported avoiding the system vendor lock through the ability of installing different devises from a verity of protocols seen from a high level as devices from the same type. It has also increased the system ability to extend its infrastructure, where new devices are automatically detected, included and updated in the high level part of the system. This is achieved while keeping a real time communication and reflection any the device event and time driven status.

The IoT concept of service oriented architecture has increased the system modularity where complex applications have been splitted to elementary modules with simply role. This has increases the system extendability, adaptability and flexibility of creating a verity of scenarios as a service composition.

Finally, having all projects, services and applications sharing the same semantic model and running on the top of the same platform has granted a full cooperation between the different entities without being syntactically or physically linked to each others. This has allowed a cooperative development add to a flexible importing and integration of other services.

After more than 6 months of operational system in real life, the implemented IoT vision within the uSmAAL system has proved a high flexibility with system integration, extensibility, adaptability, plug and Play and personalization.

6 Conclusion

Existing AAL systems are bound to already implemented services and special types of hardware devices. This lack of dynamic makes them rather unusable for the long term usage in a real life environment. Due to its characteristics, requirement and impact on real life system, the IoT has gained significant attention over the last few years. Through the made study between AAL and IoT characteristics, covered area and related requirement, we come to the conclusion that IoT systems not only fit most of AAL requirement and characteristics, but it offer also a good coverage to empower the AAL real life system chances to sustain and succeed face the dynamic and continuously requests for changes, adaption and extension add to the hardware related issues. The AAL system inspire its robustness from the implementation of the IoT characteristics, mainly from Intelligence, architecture, support of complex system, scalability, real time processing, service oriented Architecture and Middleware based implementation. The created uSmAAL system has made a successful attempt to adapt the closed AAL system to an IoT one. This experience has enriched our expertise, thus opening the door a new generation of Ambient Assisted Living solution compared to the classic closed ones. We have made an attempt through the current paper to highlight the advantages of enabling the IoT architecture for the Ambient Assisted Living system, thus assuring the system flexibility, adaptability and extendability. Further future work will mainly articulate complex situation recognition in an IoT environment, impact of real life uncertainty of data on reasoning add to empowering the explicit interaction between the IoT enabled AAL system of the end users.

References

1. Weber, R.H.: Internet of Things new security and privacy challenges. Comput. Law Secur. Rev. **26**(1), 23–30 (2010). doi:10.1016/j.clsr.2009.11.008
2. Bravo, J., Hervás, R., Villarreal, V.: Ambient assisted living: third international workshop. In: IWAAL 2011, Held at IWANN 2011, Torremolinos-Málaga, Spain, June 8–10, 2011, Proceedings, vol. 6693. Springer (2011)
3. Stengler, J., Gaikward, G., Hmida, H.B.: Towards the deployment of open platform AAL services in real life-advantages and lessons learned—usmaal: a case study for implementing intelligent AAL services in real life based on the open platform universaal. In: ICT4AgeingWell 2015—Proceedings of the 1st International Conference on Information and Communication Technologies for Ageing Well and e-Health, Lisbon, Portugal, 20–22 May, 2015, pp. 67–74 (2015). doi:10.5220/0005450200670074

4. Memon, M., Wagner, S.R., Pedersen, C.F., Aysha Beevi, F.H., Hansen, F.O.: Ambient assisted living healthcare frameworks, platforms, standards, and quality attributes. Sensors (Switzerland) **14**(3), 4312–4341 (2014)
5. Gubbi, J., Buyya, R., Marusic, S., Palaniswami, M.: Internet of Things (IoT): a vision, architectural elements, and future directions. Fut. Gener. Comput. Syst. **29**(7), 1645–1660 (2013). doi:10.1016/j.future.2013.01.010
6. For Science, T.G.O.: The internet of things: making the most of the second digital revolution (2014)
7. Dlodlo, N., Mvelase, P., Foko, T., Mathaba, S.: The state of international internet of things research. In: International Conference on Information Management and Evaluation. p. 69. Academic Conferences International Limited (2012)
8. Tan, L.: Future internet: the Internet of Things. In: 2010 3rd International Conference on Advanced Computer Theory and Engineering (ICACTE), pp. V5-376–V5-380 (2010). http://ieeexplore.ieee.org/lpdocs/epic03/wrapper.htm?arnumber=5579543
9. European Technology Platform on Smart Systems Integration: Internet of Things in 2020, pp. 1–27 (2008)
10. Saint-Exupery, A.D.: Internet of things: strategic research roadmap. Internet of Things Strategic Research Roadmap, pp. 1–50 (2009)
11. Castillejo, E., Almeida, A., López-de Ipiña, D., Chen, L.: Modeling users, context and devices for ambient assisted living environments. Sensors **14**(3), 5354–5391 (2014)
12. Alam, S., Chowdhury, M.M.R., Noll, J.: SenaaS: an event-driven sensor virtualization approach for Internet of Things cloud. In: 2010 IEEE International Conference on Networked Embedded Systems for Enterprise Applications, pp. 1–6 (2010). http://ieeexplore.ieee.org/lpdocs/epic03/wrapper.htm?arnumber=5678060
13. Patel, P., Jardosh, S., Chaudhary, S., Ranjan, P.: Context aware middleware architecture for wireless sensor network. In: 2009 IEEE International Conference on Services Computing, pp. 532–535 (2009). http://ieeexplore.ieee.org/lpdocs/epic03/wrapper.htmarnumber=5283906
14. Sundmaeker, H., Guillemin, P., Friess, P.: Vision and challenges for realising the Internet of Things (2010)
15. Banerjee, P., Friedrich, R., Bash, C., Goldsack, P., Huberman, B.A., Manley, J., Patel, C., Ranganathan, P., Veitch, A.: Everything as a service: powering the new information economy. Computer **44**, 36–43 (2011)
16. Uckelmann, D., Harrison, M., Michahelles, F.: Architecting the Internet of Things. Springer (2011)
17. Makris, P., Skoutas, D.N., Skianis, C.: A survey on context-aware mobile and wireless networking: on networking and computing environments' integration. IEEE Commun. Surv. Tutorials **15**(1), 362–386 (2013)
18. Vasseur, J.P., Dunkels, A.: Interconnecting Smart Objects with IP: The Next Internet. Morgan Kaufmann (2010)
19. Bandyopadhyay, Soma, Sengupta, Munmun, Maiti, Souvik, Dutta, Subhajit: Role of middleware for internet of things: a study. Int. J. Comput. Sci. Eng. Surv. **2**(3), 94–105 (2011)
20. Markopoulos, P.: Ambient intelligence: vision, research, and life. J. Ambient. Intell. Smart. Environ. **8**(5), 491–499 (2016). http://content.iospress.com/articles/journal-of-ambient-intelligence-and-smart-environments/ais393
21. Franchimon, F.: The new everyday: Views on ambient intelligence, by E. Aarts and S. Marzano; 2003. Gerontechnology **4**(4), 240–243 (2006)
22. Aarts, E., Wichert, R.: Ambient intelligence (2009)
23. Abdmeziem, R., Tandjaoui, D.: Internet of things: concept, building blocks, applications and challenges (2014). arXiv:1401.6877
24. Fuchsberger, V.: Ambient assisted living: elderly people's needs and how to face them. In: Proceeding of the 1st ACM International Workshop on Semantic Ambient Media Experiences, pp. 21–24 (2008). http://portal.acm.org/citation.cfm?id=1461917
25. Iliev, I., Dotsinsky, I.: Assisted living systems for elderly and disabled people: a short review assistive systems : state of art. Int. J. Bio Autom. **15**(2), 131–139 (2011)

26. Thomas, K., Becker, M., Ras, E., Holzinger, A., Müller, P.: Ambient intelligence in assisted living: enable elderly people to handle future interfaces. In: Universal Access in Human-Computer Interaction. Ambient Interaction, pp. 103–112 (2007)
27. Hanke, S., Mayer, C., Hoeftberger, O., Boos, H., Wichert, R., Tazari, M.R., Wolf, P., Furfari, F.: Universaal–an open and consolidated aal platform. In: Ambient Assisted Living, pp. 127–140. Springer (2011)
28. Liu, J.J., Wang, H.J., Zhu, Y.H.: Module for a programmable logic controller (plc) (Oct 29 2013), uS Patent D692, 397
29. Märker, M., Wolf, S., Scharf, O., Plorin, D., Teich, T.: Knx-based sensor monitoring for user activity detection in aal-environments. In: Ambient Assisted Living and Daily Activities, pp. 18–25. Springer (2014)
30. Yang, W., Lv, K., Li, M., Zhang, D.: The wireless intelligent controller of greenhouse based on zigbee. Sens. Lett. **11**(6–7), 1321–1325 (2013)
31. Tazari, M.R., Furfari, F., Fides-Valero, Á., Hanke, S., Höftberger, O., Kehagias, D., Mosmondor, M., Wichert, R., Wolf, P.: The universaal reference model for aal. Handbook Ambient Assist. Living **11**, 610–625 (2012)

Standardisation for Mobility-Related Assisted Living Solutions: From Problem Analysis to a Generic Mobility Model

Michael Brach, Armin Bremer, Andreas Kretschmer, Janina Laurila-Dürsch, Sebastian Naumann and Christoph Reiß

Abstract DKE, the German Commission for Electrical, Electronic & Information Technologies, established an AAL (ambient assisted living) standardisation roadmap and a mobility group working on a standardisation document for DIN and IEC committees. Problems in mobility support, i.e. changing body positions and locomotion, are, among others, (a) individual need changes on different time scales, (b) AAL usage as resource of both increase and decrease in activity and (c) barriers between indoor and outdoor mobility. Experts from science, technology and industry (human movement, transportation, telecommunication, computing, electrical engineering) developed (1) a common theoretical base, using existing European projects in the field of AAL, DIN ISO standards, (2) user stories on different target groups and mobility devices, (3) example use cases for the UCMR (Use Case Management Repository), (4) a generic mobility model using UML (unified modeling language). The resulting model combines action theory (situation with person, task and environment) and the resource-based view (internal and external resources) in order to distinguish mobility as room for action. Thus, standardising mobility-related AAL

M. Brach
University of Münster, Münster, Germany
e-mail: michael.brach@wwu.de

A. Bremer (✉)
serviceatmobile GmbH, Bad Endbach, Germany
e-mail: armin.bremer@serviceatmobile.de

A. Kretschmer
Beratung im Verkehrswesen, Dresden, Germany
e-mail: andreas.kretschmer@inavet.de

J. Laurila-Dürsch
DKE, Frankfurt, Germany
e-mail: janina.laurila-duersch@vde.com

S. Naumann
Institut für Automation-Kommunikation, Magdeburg, Germany
e-mail: sebastian.naumann@ifak.eu

C. Reiß
Christophorus-Consult, Offenbach, Germany
e-mail: c.reiss@christophorus-consult.de

R. Wichert and B. Mand (eds.), *Ambient Assisted Living*,
Advanced Technologies and Societal Change,
DOI 10.1007/978-3-319-52322-4_14

means to assess resources the product will offer to but also demand from the user. This holds especially when different products are combined in mobility chains.

1 Introduction

Assisted living solutions claim to support maintaining independence, which is one of the most important wishes for ageing or impaired individuals. Independence requires mobility, i.e. the capability, opportunity and performance of changing body positions and of locomotion. For mobility is a cross-sectional and complex subject, this paper deals with needs and challenges in order to support the development and interoperability of related ambient assisted living (AAL) solutions.

DKE, the German Commission for Electrical, Electronic & Information Technologies of DIN and VDE, had established an AAL standardisation roadmap [15, 16] and several working groups to continue along the roadmap and to develop standardisation documents. The working group STD 1811.0.9 Mobility [15, p. 23] includes members from transportation science and technology, human movement science, telecommunication, building industry, computer science and electrical engineering.

This diversity allowed for dealing with the topic from multiple perspectives. Starting with a problem analysis, the working group realised that the object of research and development is described in confusing terminology. Common terms and more than that, a model of mobility would be essential for effective work on interoperability.

Thus, working on common understanding and describing demands and opportunities of the topic had been the major part of the work. Interim reports were presented and/or published on Congresses in order to receive feedback from different scientific communities [5, 8, 10]. The work has reached a major milestone by finialising an application guide [9].

The present paper describes methods and proceedings of the AAL working group in order to add some background to the application guide, which is also summarised. It includes an analysis of the field mobility (Sect. 2), a common theoretical base (Sect. 3), an overview on user stories and use cases (Sect. 4) and the resulting generic mobility model (Sect. 5). Concluding remarks (Sect. 6) cover further standardization and research needs.

2 Problem Analysis

Mobility may be called *cross-sectional* from multiple points of view, because in many cases different goals and requirements interfere or are involved at the same time:

- In order to realise the wish for independent living, mobility is required in most cases.
- Mobility makes face-to-face encounters possible, which means social participation in a wholistic way.
- Mobility requires coordination of personal assistance, conventional assistance devices and technical AAL solutions.

Thus, the working group started with a problem analysis. At first, mobility-related aspects were identified from the following resources:

- Research projects funded by different calls of the German funding programme *Altersgerechte Assistenzsysteme für ein gesundes und unabhängiges Leben – AAL* (BMBF, German federal ministry of education and research),
- Research projects funded by different calls of the European *Ambient Assisted Living Joint Programme*,
- Collections of user stories or projects in the field of AAL [15, 17], and
- DIN ISO 9999 [27] (a standard including conventional mobility assistant devices).

Second, scientific overviews of research and development in transportation science, human movement science and health related projects were presented. All aspects regarding mobility were discussed and related to projects from any background disciplines. Several findings were considered key issues in the field of mobility, which should definitely be covered by any AAL solution and thus, by corresponding standardisation documents. Other findings were evaluated as potential contributions to theoretical and practical solutions of some key issues. Later in this paper, we will come back to these.

The aspects considered as key issues are summarised in Sect. 2.3. Before, we sketch some projects from public transportation (Sect. 2.1) and statements from official committees (Sect. 2.2) as a justification of these key issues.

2.1 Related Work from Public Transportation

Several research projects have been carried out in the field of public transportation with the aim improve travel information for handicapped people. The research project BAIM [28] focussed on barrier-free travel chains for people with disabilities. The follow-up project BAIMplus [1] addressed the integration of real-time data (e.g. delays) and the improvement of usability for elder people. BAIM as well as BAIMplus created a detailed path network within station buildings including stairways, escalators and lifts.

Using this approach, a public transport journey planner has been developed. Suitable routes can be found which consider the specific handicaps of individuals. For example, a handicapped person may be dependent on a lift in order to get from the entry to the platform. However, the lift may be out of order and there is no other possible route available. In this case, the public transport journey planner filters out the respective station and delivers a feasible route instead. A data structure has been created which manages user profiles. A user profile contains information like 'stairway usage yes or no', 'usage escalators yes or no', 'maximum gap between public transport vehicle and platform', 'maximum speed (walking or wheelchair)', 'maximum gradient of ramps', 'coping a step possible yes or no' or 'maximum height of the step when entering a public transport vehicle'. The work within both research projects particularly focussed the needs of wheelchair users, people with baby strollers and walking-handicapped people.

More recent research projects address the field of orientation and navigation during public transport journeys using smartphones and additional infrastructural devices like Bluetooth beacons. Thereby, the research project M4GUIDE [21] focussed on blind people, whereas the research project NAMO [11] aimed to support elder travellers. Within the research project DYNAMO [12], a navigation system for station buildings has been developed which bases on Bluetooth beacons and pedestrian dead recognising (PDR) aiming at healthy (non-handicapped) people.

Compared to this related work, our paper proposes a more generic approach of a mobility model explicitely supported by AAL solutions and including mobility inside the flat. Combining the skills of people with possible barriers on the path network (e.g. stairway) enables public transport journey planning and navigation systems to find suitable routes for disabled people and to help carrying out the journey. Our approach also deals with skills and properties of route sections, however, in much more detail by including knowledge of human movement sciences.

2.2 Related Work from Official Committees

In the context of the *German 2015 National IT Summit* in Berlin, a focus group on smart data and intelligent transport systems discussed the interplay of informational and physical issues in transport systems, e.g. to integrate sensor-related data with existing information on mobility and transport infrastructure [23, p. 4]. Describing user stories and use cases (e.g., travelling across Germany using different transport systems like tram, bus, railway, bicycle), the authors state [23, p. 33]: "*Transitions between different routing networks are always challenging: where are the access points? How are they connected with the road system? Are there business hours?*"[1]

In 2009, the *German Federal Ministry of Education and Research* (BMBF) implemented an expert panel focussing on non-technical aspects of AAL and contributing to accompanying research from an independent and higher-level perspective. In 2011, the *Loccum Memorandum: Technical Assistance Systems for Demographic Change—an Inter-generational Innovation Strategy* [18, 19] was published.

[1]Original text translated by the authors of this paper: "Eine Herausforderung sind immer die Übergänge zwischen verschiedenen Routingnetzen; Wo sind Zugänge, wie sind diese mit dem Straßennetz verknüpft; gibt es Öffnungszeiten?" [23, p. 33].

Among other recommendations, the expert panel postulates, that AAL solutions should facilitate and utilise existing potentials and resources and that the users' practical knowledge should be included. The panel considers not only elderly persons, but young and old, healthy and ill and handicapped persona as users of assistive technical systems. Especially, stigmatisation should be avoided.

In consequence, such systems have to be scalable to very different user needs and modular approaches should be preferred. In this paper, we follow these recommendations from the technical and the non-technical perspective as well. The following subsection will summarise the requirements.

2.3 Key Issues in the Field of Mobility

As a result of the procedure of collecting and discussing aspects from different disciplines, and from technical and non-technical perspectives,the following issues were identified as being critical in the field of mobility. Assisted living solutions and thus the targeted application guide should deal with these:

Informational and physical components. Need for assistance and supply of assistance includes information and orientation (e.g. routing, signs) as well as physical support (e.g. walking stick, car). The components may stand alone or may be combined.

Intentions and resources. Users have their own intentions and motives. They make decisions and plans. Different individuals will deploy their resources differently, but also evaluate things differently. For instance, saving time, saving effort, or mastering a challenge are individual criteria. Assistive solutions should allow for tailoring to individual favours.

Dynamic assistance needs. The individual need for support changes on different time scales, due to daily fitness, health or ageing processes. Also training processes due to learning (cognition) or biological adaptation (physical) may be considered. Assistive offers should follow changing needs in either direction and configuration.

Stimulating character. The usage of assisted living utilities itself can modify the user's mobility in a good or harmful way, similar to conventional mobility devices. Up to now, new technical devices are used for increasing comfort in travelling (e.g., car) or finding one's way (e.g., navigation system). As a consequence, previous individual abilities tend to decrease (e.g., distance walking, reading maps). In the mean time, combinations of sensors and actors became typical for mobility solutions, especially in the context of AAL. Therefore, more than just being used for a fixed supporting task, AAL solutions can offer choices in a way that existing abilities are utilised and secured. More than that, cognitive and physical mobility skills can be facilitated and exercised in general. AAL should utilize these opportunities for stimulation.

Indoor outdoor transition. Users of mobility support often experience disruption between indoor and outdoor mobility, for devices have to be changed or don't work in both areas. This should be considered an improved in assisted living solutions.

Conclusion. Assistance solutions in the field of mobility should cover these key issues, or at least should consider them and be prepared for later combination and interoperability. Consequently, in order to continue standardisation work in this field, also the theoretical framework of mobility and the mobility model should cover these characteristics.

3 Theoretical Framework

In this section, mobility will be conceptualised twofold: by a discussion of the term mobility and by a description of a framework based on two theories.

In the first perspective, we define mobility as changing the position of the body and/or locomotion in the most generic way. For example, turning from the back to a side position, while lying in the bed, is covered by the term mobility. One can think about personal, conventional and AAL solutions in order to assist with this transition. This is also an example for mobility within the home. The environment is an important moderator for mobility, but there is no restriction regarding the setting, where mobility takes place. Locomotion and position changes may happen within a car seat (e.g., turning the body to position the feet near the pedals), within a bed, indoor or outdoor, inside means of public transport and so on.

However, we delineate our concept of mobility from similar terms, such as social mobility (change of social status), residential mobility (moving to a new home), cognitive mobility (flexible thinking as a conception, while cognition in general is required for mobility action), transportation of goods and freight (with the exception, that the individual may have to carry some luggage or uses a mobility device).

An important dimension is potential mobility versus its factual performance. Both aspects are covered. The first case incorporates an option to use transportation services and any other assistance or to rely on transportation and mobility infrastructure (e.g. road, stairs, door). In most cases there are different choices of mobility: going by feet, using a wheeler, a car and so on. The individual makes a decision and configures the transfer from A to B. This includes the choice not to start a move or transfer at all. The second case describes the actual performance of a mobility action. This concrete action is one of the possible implementations of the potential mobility. Per definitionem, the potential mobility sets limitations to the factual activities. These may or may not not follow the planned action.

However, the direction of influence reverses on mid and long terms. Reducing factual mobility can reduce mobility potentials by means of breaking habits (e.g., walking regularly) and developing fears (e.g. fear of falling) and biological adaptation laws. On the other hand, intelligent AAL solutions may increase factual mobility, and on mid and long terms may help to build new habits, reduce fears and thus biological adaptation again may extend mobility limitations. The same holds for transportation and mobility infrastructure.

In consequence, both the potential and the actual mobility are important, because both contribute to social participation, self-determined life, health and well-being.

In the second perspective, the working group combined two theories in order to model potential and factual mobility as described above and to consider the key

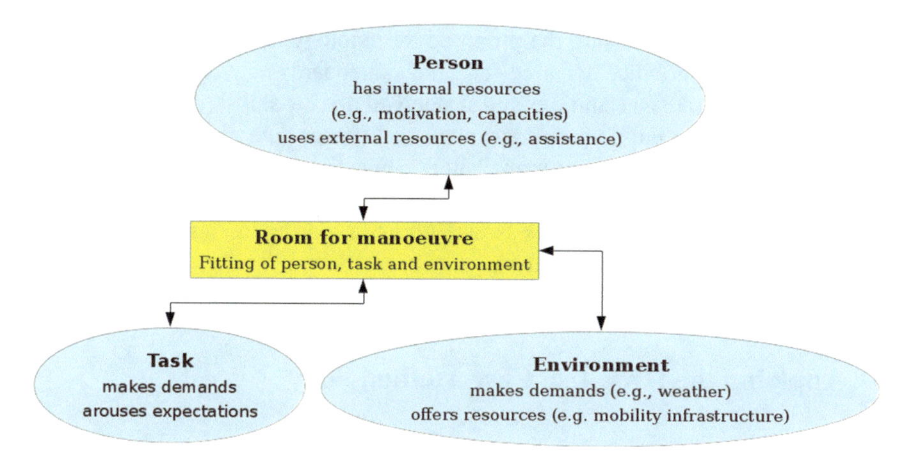

Fig. 1 Mobility as room for manoeuvre: a framework from action theory and resource view

issues summarised in Sect. 2.3. The general idea is to describe potential mobility as room for manoeuvre, as a space for action, which is spanned by the person, the mobility task and the environment (see Fig. 1).

Nitsch and colleagues introduced action theory [24] into sport-related actions from a psychological point of view, and defined a situation by three components: an individual, a task, an environment. The person comes with abilities, skills, and motivational dispositions. The terminology of abilities and skills are will be described in the generic model in Sect. 5. The task includes certain demands, which will be evaluated by the person and will cause expectations and motivation according to its incentive structure (e.g., difficulty and competence, value and importance). The environment modulates the person–task–relation, e.g. some mobility tasks may be less or more difficult depending on the weather. Decisions and actions result from these components. The framework has proved to fit also to physical activity in general, a later description and discussion is found in [25].

For the purpose of the standardisation work covered in this paper, we combined the framework with the so-called *resource view*. This theory originally came from work physiology and has been has been conditioned for sport and exercise by Schönpflug [26].

We use it here for streamlining language. An example will show, that certain features of the three situation components share the same nature: In order to reach the next bus at the downhill station (task), one could use a personal capacity (running), a mobility device (skateboard) or an environmental mobility infrastructure (short cut by stairs). Alternatively, the person may decide to change the task demands and go for the following bus 20 min later. All these are resources, which are internal or external to the person and can complement each other. All possible actions which fit the situation (person, task, environment) are part of the room for manoeuvre, in other words, the potential mobility.

Two other benefits of using the proposed terminology is, that (a) the benefit of a certain external mobility resource (e.g. the skateboard) may depend on certain internal resources (balance abilities and skateboard driving skills), and (b) resources may be supplied and build up or consumed on different time scales. For instance, using a car accelerates mobility immediately (short time scale), but may influence walking mobility after years (long time scale).

Thus, the theoretical framework is useful in considering the issues which have been described in the problem analysis.

4 Applying the DKE Use Case Methodology

After agreeing the theoretical framework, the working group followed the initial steps of the official DKE standardisation methodology [13, 14], in order to develop the generic model of mobility and the application guide:

1. Developing complex user stories, covering different target groups and mobility devices
2. Developing example use cases, using the official UCMR format (Use Case Management Repository) from standardisation procedures
3. Developing the generic mobility model, using UML (unified modelling language)
4. Describing example use cases in UML

Table 1 Features of the user stories

Feature	User story 1	User story 2	User story 3	User story 4
Main character	Arthur	Ute & Bernd	Tom	Marla
Target age group	Very old	Varying	Old	Young adult
Point of view	End user	Provider	End user	End user
Means of transport	Pedestrian	Bicycle	Car	Infrastructure
Informational and physical components	X	X	X	X
Intentions and resources	X	X	X	X
Dynamic assistance needs	X	X	X	X
Stimulating character	X		O	X
Indoor outdoor transition	X	O		

Symbols: X—feature is covered, O—feature is partly covered

The Table 1 presents major features of the user stories, which were developed for this purpose and for use in the application guide [9]. All key issues (Sect. 2.3) are covered in several stories, and different perspectives are included. Uses cases were identified from the stories. The generic mobility model has been applied to the stories. These steps are not part of the present paper. The user stories were collected in a working paper [6] for further analysis.

5 A Generic Mobility Model

The mobility model assumes that mobility actions are performed by individuals. The term 'mobility action' includes not only outdoor trips (e.g. shopping or doctor including different modes of transport) but also trips within the flat (e.g. from the table to the fridge or to the bed). A mobility action can also be considered as a section of a route.

Resources play a central role in the mobility model. They are required to perform mobility actions. Technically spoken, a mobility action knows which resources are required in order to perform itself. A person has resources for performing mobility actions. AAL solutions immediately compensate missing or poorly trained resources. Additionally, they improve resources in mid-term and long-term by training. On the other hand, using AAL solutions requires consumptive resources and, furthermore, may have disadvantages (e.g. coping a stairway with a rollator).

Figure 2 shows the mobility model as UML class diagram concerning the immediate impact of an AAL solution. Mobility action, resource, person and AAL solution are defined as classes. Concrete objects can be created from such classes. Such concrete objects can have different properties even if they belong to the same class. These different properties are described by class attributes. The connections between the classes are to read as 'a mobility action requires resources', 'a person has resources',

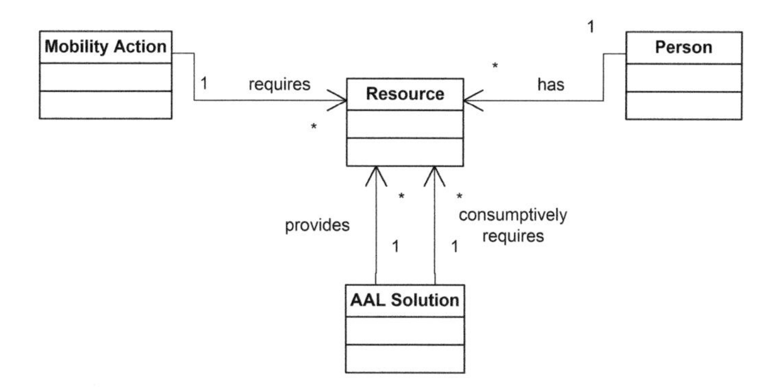

Fig. 2 Mobility model

'an AAL solution provides resources' and 'an AAL solution consumptively requires resources'.

A (composite) mobility action should be divided into a sequenc of elementary mobility actions. Each elementary mobility action requires certain resources. As with the skills, elementary mobility actions should be described in detail by attributes. A mobility action may require several resources, the exact number is not fixed.

AAL solutions immediately compensate missing or poor trained skills and abilities of people to deal with mobility actions. Beside this, AAL solutions can also contribute to targeted training effects, because in the course of using the AAL solution consumptive resources are used. If an over-threshold stimulus is present, structural resources are built.

5.1 Resources

Resources can be considered as *abilities* or as *skills*. The main sensorimotor *abilities* needed for mobility are:

- *Strength*: the ability to change muscle tension in order to perform concentric and eccentric work and to assume and to maintain body positions and postures;
- *Balance*: the ability to obtain or regain equilibrium by interaction of the various senses, reflexes and motor control functions;
- *Orientiation*: complex cognitive and sensorimotor ability to detect targets and directions to align body posture and locomotion in order to identify and select options for action.

The three abilities mentioned are important for mobility, but do not cover all requirements. Other abilities would be conceivable, but are not considered here. Each of these abilities includes related cognitive aspects concerning the recording and processing of information as well as decision-making and movement regulation processes.

Research in sport and exercise science shows that limitations of the ability approach become more evident the more specific and powerful movement actions are considered [20, 22].

In addition to considering general abilities, another approach based on *skills* has proven. Here, concrete mobility actions are considered, and specific requirements are determined. For example, climbing a staircase requires the one-legged lifting or lowering of body weight for each step. The use of a signaled road crossing requires a certain walking speed. The corresponding resources are called sensorimotor skills. They include the necessary perception and muscle functions to perform the corresponding action. Specific skills are described in more detail by attributes. For example, the skill 'going forward' owns an attribute 'distance', describing how far a person can go forward depending on its day's constitution.

5.1.1 Sensorimotor Skills

Each elementary mobility action requires specific skills. In Table 2, such elementary mobility actions are exemplarily compiled (see [2, 3, 7]) and sorted into three groups within the meaning of functional units:

- *Posture*: Although it may seem like a contradiction at first glance, holding a body position is necessary for mobility, e.g. while waiting for an elevator. Examples are holding a sitting posture (30 min) and standing (60 s).
- *Transition*: A transition is an acyclic change between two body positions or forms of locomotion. Examples for transitions from a lying position are from supine to side position and from lying on the side to sitting.
- *Locomotion*: The locomotion includes locomotion in a strict sense. When walking forward we distinguish short distances (e.g. 6 m within a room or entering a vehi-

Table 2 Elementary mobility actions with required resources

Elementary mobility action	Required resources
Bus trip	Sitting Standing
	Transition from standing to sitting
	Transition from sitting to standing
	Going up stairs
	Going down stairs
	Visual perception
	Going forward
	Transition from standing to going
	Transition from going to standing
Plane path	Going forward
	Transition from standing to going
	Transition from going to standing
Ramp	Going forward
	Transition from standing to going
	Transition from going to standing
Road crossing	Going forward
	Transition from standing to going
	Transition from going to standing
	Visual perception
Steps	Going up stairs
	Going down stairs
Door	Pushing down doorknob
	Pulling door
	Pushing door

cle) and longer distances (e.g. 60 m for paths on a floor, entrances and exits for transport) because different combinations of external and internal resources may be required.

Two more skills from the mentioned motor tests [3, 7], may exemplary be named. These do not belong to mobility in the narrower sense, however, they can get a functional significance, for example in the use of aids. Examples are grabbing to the top while standing or picking a subject from the ground.

5.1.2 Cognitive Skills

In addition to sensorimotor skills, cognitive skills of perception and orientation are relevant for mobility actions:

- Visual perception (recognition, reading, understanding of signs);
- Locomotor control skills (holding a direction to move on to something or from something away, keeping distances and change tracking skills);
- Simple reactions and selection reactions (decisions).

However, so far mobility-related cognitive skills are less concretized than sensorimotor skills.

5.2 Application

If all data of the mobility actions, persons and AAL solutions have been completely collected, the following questions can be automatically answered based on the mobility model:

- *Calculation of the compensation needs*: For a given mobility action and a given person the model can be used to check whether all the required resources of the mobility action are met by this person. A result in the negative case could be a list of required but actually not covered resources;
- *Determination of appropriate AAL solutions*: Existing AAL solutions could be selected based on the previously determined compensation requirements;
- *Determination of the mobility action depending on the available resources*: For a given start location, a given destination location and the resources of a person (personal resources including AAL solutions) an appropriate mobility action is searched the person is able to perform.

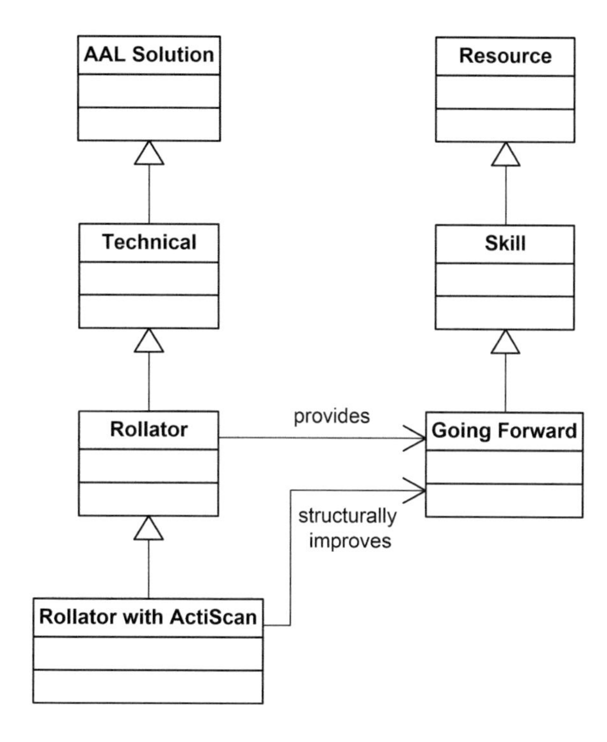

Fig. 3 Immediate provision of the resource 'going forward' by the rollator and medium- to long-term improvement through the guided training of an AAL product

5.3 *Example*

During this section, the mobility action 'zoo visit' along with the 'ActiScan' function of an intelligent rollator is exemplary modeled. Arthur's walking distance is severely limited by osteoarthritis in knees and hips. Actually, Arthur is not able to cover more than 250 m without relaxing break. The rollator helps him to increase this distance immediately. The 'ActiScan' function of the rollator motivates Arthur to increase walking training, which results in a structural improvement of the resource 'going forward' in medium and long term (see UML class diagram in Fig. 3).

For modeling the mobility action 'zoo visit', some possible elementary mobility actions are provided in the UML class diagram in Fig. 4.

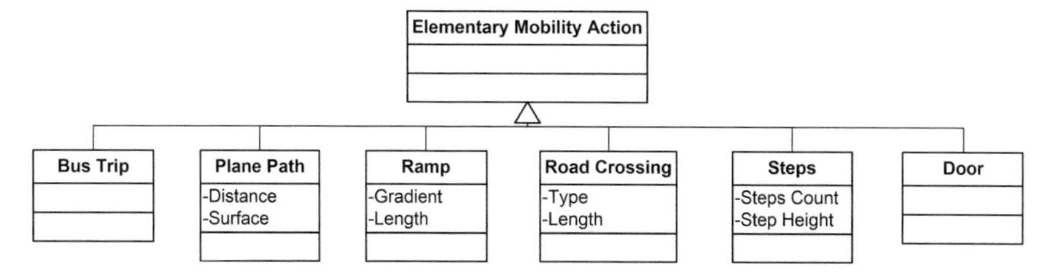

Fig. 4 Possible elementary mobility actions

The resources required by each elementary mobility action are shown in Table 2 and not in another UML class diagram, in order to use a more compact form. The concrete mobility action 'zoo visit' arises from the sequence of the elementary mobility actions.

Except for the limited walking distance (resource 'going forward'), Arthur has no limitations in sensorimotor and cognitive skills. The limitation on the resource 'going forward' means that Arthur is not able to carry out the whole mobility action 'zoo visit'. However, another 'zoo visit' mobility action, which Arthur could perform independently, does not exist. Only the use of a rollator enables Arthur to visit the zoo. The shortest walking route leads through a staircase with 10 steps that Arthur can not cope with the rollator. The rollator therefore proposes another route via a ramp instead.

6 Discussion

Starting from a heuristic analysis of projects and texts in the field of ambient assisted living, the standardisation working group on mobility presented a problem analysis with key issues, a theoretical conception of mobility and a generic mobility model. Numerous practical examples were given in order to come to a common understanding within the group and with the reader of the present paper.

From a more formal point of view, the first steps of the official DKE methodology have been taken. The result is an application guide [9], which has successfully been subject to commenting in the DKE committees and will be published in a few months.

The main strength of the generic UML model is, that it has been built on a theoretical base from action psychology and from work physiology. This theoretical background has proven useful in human movement science, which includes not only sport but also everyday physical activity. Thus, user behaviour is included not only as "movements" (a more technical viewpoint) but as series of intended, motivated actions which apply, consume and build internal and external resources on different time scales, and thus influence future actions in a positive or negative way. Therefore, the theory and thus the practical generic model includes not only a "user", but a human being. This strength is innovative in the field of AAL and may be useful also for other AAL subjects.

On the other hand, the main weakness of the results presented in this paper is that up to now there are only examples and no real applications. Thus, feedback is needed in order to identify critical points and to work on them. More use cases have to be elaborated (see the application guide for some examples [9]). The next steps in the DKE methodology should be taken.

The applications stated in Sect. 5.2 and the elementary mobility actions should be studied from the scientific point of view. Practical applications are the definition of quality criteria for AAL-ready mobility infrastructure or certified AAL products. The basic idea of such a quality (Q): Each conventional device, infrastructure or technical solution should describe, (Q1) which resources are provided and (Q2) which resources are required, when the respective device is used. The description should consider (Q3) short time and (Q4) long time scales.

The model presented here may be a starting point for such a description.

References

1. Becker, J., Pilz, A., Twele, H., Heck, H.: Mobilität durch Information: Forschungsprojekt BAIM plus—Bilanz und Ausblick. Mobility thanks to information. NAHVERKEHR **29**(4) (2011)
2. Brach, M.: MOTA – ein Mobilitätstest für alte Menschen. Ergotherapie und Rehabilitation **37**, 104–107 (1997). https://www.researchgate.net/publication/233882805_MOTA_-_ein_Mobilittstest_fr_alte_Menschen. Cited 27 Nov 2015
3. Brach, M.: Motor milestones for elderly people in institutional care: adapting MOVE to a new target group. In: The Disability Partnership (ed.) Conference 2000 "Moving Towards Inclusion", Chatsworth House, Derbyshire. Conference Proceedings Document, pp. 15–17. The Disability Partnership, London (2000). doi:10.13140/RG.2.1.2064.0087
4. Brach, M.: Research on exercise programs-an approach of technological science. Eur. Rev. Aging Phys. Act. **6**(2), 63–65 (2009). doi:10.1007/s11556-009-0053-x
5. Brach, M., Korn, O.: Assistive technologies at home and in the workplace–a field of research for exercise science and human movement science. Eur. Rev. Aging Phys. Act. **9**(1), 1–4 (2013). doi:10.1007/s11556-012-0099-z
6. Brach, M., Kretschmer, A.: Mobilitätsbezogene User Stories. Arbeitspapier für den Arbeitskreis AK 1811.0.9 Mobilität, VDE Verband der Elektrotechnik Elektronik Informationstechnik e.V., VDE, Frankfurt (2015). doi:10.13140/RG.2.1.3010.3929
7. Brach, M., Haasenritter, J., Kirchner, E., Bauder-Mißbach, H., Betschon, E., Eisenschink, A.M., Drabner, A., Panfil, E.-M.: Painful movements and mobility after urological surgery: studying the feasibility of pre-operative exercise, a new mobility test and a randomised controlled trial protocol with cystectomy patients in intensive care [ISRCTN32898285]. Webmed-Central NURSING 3: article no. WMC003102 (2012)
8. Brach, M., Reiß, C., Boos, H., Kretschmer, A., Naumann, S., Bremer, A., Laurila-Dürsch, J.: Zuhause und unterwegs: Normen unterstützt die Entwicklung alltagsgerechter Assistenzlösungen im Themenfeld Mobilität. Conference: Wohnen – Pflege – Teilhabe "Besser leben durch Technik": 7. Deutscher AAL-Kongress mit Ausstellung, 21. - 22. Januar 2014, Berlin (2014). doi:10.13140/RG.2.1.4329.4483
9. Brach, M., Bremer, A., Kretschmer, A., Laurila-Dürsch, J., Naumann, S., Reiß, C.: Technikunterstütztes Leben – Ambient Assisted Living (AAL) – Mobilität im AAL-Umfeld. VDE Anwendungsregel VDE-AR-E 2757-9 des AK 1811.0.9 Mobilität, VDE Verband der Elektrotechnik Elektronik Informationstechnik e.V., VDE, Frankfurt (in press)
10. Brach, M., Bremer, A., Kretschmer, A., Laurila-Dürsch, J., Naumann, S., Reiß: A generic mobility model for standardisation of mobility-related assisted living solutions: a contribution from human movement science. In Baca A (ed) Crossing Borders through Sport Science – 21st Annual Congress of the European College of Sport Science. Abstracts, European College of Sport Science, Vienna (accepted)
11. Bühler, C., Heck, H., Nietzio, A., Reins, F., Berker, F.: Way-finding support in public transport environments provided by the NAMO mobile travel assistance system. Stud. Health Technol. Inform. **217**, 461 (2015)
12. Czogalla, O., Naumann, S.: Pedestrian guidance for public transport users in indoor stations using smartphones. In: 2015 IEEE 18th International Conference on Intelligent Transportation Systems (ITSC), pp. 2539–2544 (2005)

13. DKE Deutsche Kommission Elektrotechnik Elektronik Informationstechnik in DIN und VDE (ed): Die Use Case Methode. https://www.dke.de/de/std/Documents/UseCaseMethode.pdf. Cited 21 Nov 2015
14. DKE Deutsche Kommission Elektrotechnik Elektronik Informationstechnik in DIN und VDE (ed.): Use Case Management Repository. https://www.dke.de/de/std/Seiten/UseCaseManagementRepository.aspx. Cited 21 Nov 2015
15. DKE German Commission for Electrical, Electronic & Information Technologies of DIN and VDE: The German AAL Standardization Roadmap (=Ambient Assisted Living). VDE, Frankfurt (2012). http://www.dke.de/de/std/AAL/Seiten/AAL-NR.aspx. Cited 4 Dec 2013
16. DKE German Commission for Electrical, Electronic & Information Technologies of DIN and VDE: The German Standardization Roadmap AAL (Ambient Assisted Living). Status, Trends and Prospects for Standardization in the AAL Environment. Version 2. VDE, Frankfurt (2014). https://www.vde.com/en/dke/std/documents/nr_aal_en_v2.pdf. Cited 27 Feb 2016
17. Gaßner, K., Conrad, M.: ICT enabled independent living for eldery. A status-quo analysis on products and the research landscape in the field of Ambient Assisted Living (AAL) in EU-27. Institute for Innovation und Technology, Berlin (2010). http://www.vdivde-it.de/publications/studies/ict-enabled-independent-living-for-elderly.-a-status-quo-analysis-on-products-and-the-research-landscape-in-the-field-of-ambient-assisted-living-aal-in-eu-27/at_download/pdf. Cited 05 May 2016
18. Gothe, H., Grunwald, A., Hackler, E., Meyer, S., Mollenkopf, H., Niederlag, W., Rienhoff, O., Steinhagen-Thiessen, E., Szymkowiak, C.: Loccumer Memorandum: Technische Assistenzsysteme für den demographischen Wandel—eine generationenübergreifende Innovationsstrategie. Empfehlungen des Expertenrats (2011). http://www.vdivde-it.de/publikationen/leitfaeden/loccumer-memorandum-technische-assistenzsysteme-fuer-den-demographischen-wandel-2013-eine-generationenuebergreifende-innovationsstrategie/at_download/pdf. Cited 07 May 2016
19. Gothe, H., Grunwald, A., Hackler, E., Meyer, S., Mollenkopf, H., Niederlag, W., Rienhoff, O., Steinhagen-Thiessen, E., Szymkowiak, C.: Loccum memorandum: technical assistance systems for demographic change—an inter-generational innovation strategy. Recommendations of the Expert's Council (2011). http://www.health-academy.org/ha/_data/2011_01_11_Engl-AAL-Loccumer_Memorandum_web.pdf. Cited 18 May 2016
20. Jendrusch, G., Brach, M.: Sinnesleistung im Sport. In: Mechling, H., Munzert, J. (eds.) Handbuch Bewegungswissenschaft - Bewegungslehre, pp. 175–196. Hofmann, Schorndorf (2003)
21. m4guide – mobile multi-modal mobility guide. http://www.m4guide.de. Cited 17th May 2016
22. Mechling, H.: Von koordinativen Fähigkeiten zum Strategie-Adaptionsansatz. In: Mechling, H., Munzert, J. (eds.) Handbuch Bewegungswissenschaft - Bewegungslehre, pp. 347–369. Hofmann, Schorndorf (2003)
23. Nationaler IT-Gipfel Berlin: Smart Data für intelligente Mobilität. Ergebnisdokument der Fokusgruppe Smart Data für intelligente Mobilität. Plattform "Digitale Netze und Mobilität". Nationaler IT-Gipfel Berlin, Oktober 2015 (2015). http://www.bmvi.de/SharedDocs/DE/Anlage/Digitales/it-gipfel-fg-smartdata.html?nn=38976. Cited 17th May 2016
24. Nitsch, J.R.: Zur handlungspsychologischen Grundlegung der Sportpsychologie. In: Gabler, H., Nitsch, J.H., Singer, R. (eds.) Einführung in die Sportpsychologie, pp. 188–271. Hofmann, Schorndorf (1986)
25. Nitsch, J.R.: Motivation reconsidered—an action-logical approach. In: Stelter, R., Roessler, K.K. (eds.) New Approaches to Sport and Exercise Psychology, pp. 55–82. Meyer & Meyer Sport, Oxford (2005). https://www.researchgate.net/publication/301821172_New_Approaches_to_Sport_and_Exercise_Psychology?ev=prf_pub. Cited 8th May 2016. ISBN 1-84326-149-1
26. Schönpflug, W.: Beanspruchung und Belastung bei der Arbeit - Konzepte und Theorien. In: Kleinbeck, U., Rutenfranz, J. (eds.) Arbeitspsychologie. Enzyklopädie der Psychologie, Themenbereich D, Praxisgebiete, Serie III, Wirtschafts-, Organisations- und Arbeitspsychologie, vol. 1, pp. 130–184. Hogrefe, Göttingen (1987)

27. Standard ISO 9999:2011: Assistive products for persons with disability—classification and terminology. German version EN ISO 9999:2011 (2011)
28. Wahlster, M.N., Becker, J., Pilz, A., von Grumbkow, P.: Informationsdienste für mobilitätseingeschränkte Menschen Der Nahverkehr, Heft 11/2007. Alba-Verlag, Düsseldorf (2007)

Printed in the United States
By Bookmasters